Semiotic Engineering Methods for Scientific Research in HCI

Synthesis Lectures on Human-Centered Informatics

Editor

John M. Carroll, Edward M. Frymoyer Professor of Information Sciences and Technology, Penn State University

Common Ground in Electronically Mediated Communication
Andrew Monk
University of York, U.K.

Studies of Work and the Workplace in HCI
Graham Button, Sheffield Hallam University, and
Wes Sharrock, The University of Manchester

Semiotic Engineering Methods for Scientific Research in HCI
Clarisse Sieckenius de Souza and Carla Faria Leitão
Pontifical Catholic University of Rio de Janeiro (PUC-Rio)

Information Cultivation for Personal Information Management
Steve Whittaker
University of Sheffield, U.K.

Semiotic Engineering Methods for Scientific Research in HCI
Clarisse Sieckenius de Souza and Carla Faria Leitão
www.morganclaypool.com

ISBN: 9781598299441 paperback

ISBN: 9781598299458 ebook

DOI: 10.2200/S00173ED1V01Y200901HCI002

A Publication in the Morgan & Claypool Publishers series

SYNTHESIS LECTURES ON HUMAN-CENTERED INFORMATICS # 2

Lecture #2

Series Editor: John M. Carroll, Pennsylvania State University

Series ISSN

ISSN 1946-7680 print

ISSN 1946-7699 electronic

Semiotic Engineering Methods for Scientific Research in HCI

Clarisse Sieckenius de Souza and Carla Faria Leitão
Pontifical Catholic University of Rio de Janeiro

SYNTHESIS LECTURES ON HUMAN-CENTERED INFORMATICS # 2

MORGAN & CLAYPOOL PUBLISHERS

ABSTRACT

Semiotic engineering was originally proposed as a semiotic approach to designing user interface languages. Over the years, with research done at the Department of Informatics of the Pontifical Catholic University of Rio de Janeiro, it evolved into a semiotic theory of human–computer interaction (HCI). It views HCI as computer-mediated communication between designers and users at interaction time. The system speaks for its designers in various types of conversations specified at design time. These conversations communicate the designers' understanding of who the users are, what they know the users want or need to do, in which preferred ways, and why. The designers' message to users includes even the interactive language in which users will have to communicate back with the system in order to achieve their specific goals. Hence, the process is, in fact, one of *communication about communication*, or metacommunication. Semiotic engineering has two methods to evaluate the quality of metacommunication in HCI: the semiotic inspection method (SIM) and the communicability evaluation method (CEM). Up to now, they have been mainly used and discussed in technical contexts, focusing on how to detect problems and how to improve the metacommunication of specific systems. In this book, Clarisse de Souza and Carla Leitão discuss how SIM and CEM, which are both qualitative methods, can also be used in scientific contexts to generate new knowledge about HCI. The discussion goes into deep considerations about scientific methodology, calling the reader's attention to the essence of qualitative methods in research and the kinds of results they can produce. To illustrate their points, the authors present an extensive case study with a free open-source digital audio editor called Audacity. They show how the results obtained with a triangulation of SIM and CEM point at new research avenues not only for semiotic engineering and HCI but also for other areas of computer science such as software engineering and programming.

KEYWORDS

semiotic engineering, HCI theories, qualitative methods in HCI, communicability evaluation, semiotic inspection

Preface

The first comprehensive presentation of semiotic engineering as a *semiotic theory of human–computer interaction* (HCI) appeared only a few years ago (de Souza, 2005), although our research is almost two decades old now. Like all other HCI theories and approaches, our aim is to advance knowledge about HCI and thus help professionals to design and develop computer technologies that are more usable, useful, enjoyable, and empowering. However, semiotic engineering has certain unique characteristics that may play an important role in times when *thinking out of the box* is a valued practice for innovation.

The foundational distinction proposed by the theory is that HCI is not really about how users interact with computers, but rather about how users communicate with computer system designers and developers through their proxy at interaction time, which we call *the designers' deputy*. This represents an important shift in framing HCI design problems, compared to the still dominant user-centered design paradigm. Don Norman's (n.d., 2007) own reading of semiotic engineering is probably the best expression of such shift:

> It is common to think of interaction between a person and technology as communicating with the technology. De Souza shows that the real communication is between designer and person, where the technology is the medium. Once designs are thought of as shared communication and technologies as media, the entire design philosophy changes radically, but in a positive and constructive way. (Norman, n.d.)

Although semiotic engineering is *not* semiotics, HCI researchers and practitioners need to learn some basic semiotic concepts in order to understand and assess the new possibilities brought about by this perspective. These concepts have been carefully defined, contextualized to the HCI domain, and illustrated in our previous publication (de Souza, 2005). In this book, we include a brief glossary of terms to help first-time semiotic engineering readers follow our presentations and discussions. We hope that this will facilitate a shift in perspective and subsequent judgments about the value of our contribution.

Viewing HCI as *shared communications between users and designers* implies that interaction designers have an explicit role in this theory's models, namely, the same as that of users and systems. Designers are *communicators* at interaction time. They are thus along with system and users, participants in a special kind of computer-mediated communication. This communication is not *natural*. For example, natural language, which is used in all human conversations, is replaced with other kinds of signs like interface buttons, menus, images, sounds, used in combination with various sorts of input/output (I/O) devices. Gestures and facial expressions, and even speech, when present in the designer–user communication process, must be generated and/or interpreted by a *third party*, which mediates the process and acts out the designers' part before users, namely, the system.

We have often been asked why we decided to go through all the trouble of postulating that systems represent designers at interaction time. What is wrong with just saying that users communicate with systems and vice versa? The short answer is: because computer systems expose a wide range of psychological states and behavior that are nothing less but *human*. Even in the most *nonhuman* of guises, like that of a spreadsheet, the *system* will every once in a while turn to the user and say things like "Do you want to save changes you made?" or "The system cannot determine which rows or columns contain labels."

Although the above is only the short answer to why semiotic engineering models HCI as computer-mediated designer–user communication, the two system messages (extracted from dialogues with a popular commercial product) show that the users' inferences about whom *they are talking to* (or who *is talking to them*) are likely to stumble on riddles. It is fine to suppose that *the system* is asking if they want to save their file. But why would the system refer to *itself* as "the system" when saying that *it* cannot determine which rows and columns contain labels? Somebody else must be speaking, but who? Ambiguities like these are abundant in every computer system, and the disorientation they cause in the users' minds, when users try to anticipate and understand computer behavior, should not be underestimated.

The long answer to why it is advantageous to view systems as *the designers' deputies* is what this book is about. It reviews the main concepts of semiotic engineering and presents an in-depth description of the *semiotic inspection* and the *communicability evaluation* methods. Both were used in an extensive case study with an existing system, whose results illustrate some of the main contributions of the theory for contemporary scientific research and reflective professional practice.

Unlike the 2005 book, which was broader in scope and aimed at a wider readership, this book has been written for researchers, students, and professionals interested in learning more about how methods and theories support and improve each other, especially in qualitative research. In the following pages, the reader will not only learn about the theory and its methods but also orbit around

HCI on a very different path. It is our hope that this will be, in itself, a stimulating and rewarding intellectual experience, which will bear fruit in the near and not-so-near future.

We thoroughly enjoyed writing this monograph and learned a new lesson with almost every page. We wish our readers will share much of this experience.

Rio de Janeiro, summer of 2009.

C.S.S. and C.F.L.

Contents

CHAPTER 1

Introduction

In recent years, there have been repetitive calls for innovation, with increased awareness of how important it is not only to be an innovator but also to stimulate and support innovation by others. In an article in the *Communications of the ACM*, Shneiderman (2007) underlines the importance of creativity support tools, which should "support discovery in the sciences, exploration in design, innovation in engineering, and imagination in the arts" (p. 21). Although he is mostly speaking about computer tools, Shneiderman also notes the importance of methodological tools in achieving the proposed goal. He says "researchers who study and evaluate software usage are getting past old strategies of controlled studies and short-term usability testing to embrace ethnographic styles of observation, long-term case studies, and data logging to understand patterns of usage" (p. 25).

A related view has been expressed by Greenberg and Buxton (2008), who fear that, some of the time, focusing on usability can be harmful. According to the authors, usability evaluation has become the norm. It is enforced by educational programs, academic review processes, and user advocacy groups. This is certainly due its success in identifying usability problems of interactive systems, contributing to a vast body of research and professional knowledge. However, they say, many times usability evaluation is carried out "by rule rather than by thought" (p. 111). The risk is that there are situations where this method can prematurely suppress innovative ideas. Highly innovative technologies are immature by definition, and interaction with it is likely to have its blunders. Because usability evaluation tends to put the lens on blunders, and not on the big picture, innovators are likely to be discouraged by negative results and give up on plans that might otherwise bear good fruit. Likewise, authors say, usability evaluation should be used with caution to validate scientific research results. In their words, "it may incorrectly suggest a design's scientific worthiness rather than offer a meaningful critique of how it would be adopted and used in everyday practice" (p. 111).

Compared to usability evaluation, the methods presented and discussed in this book—the semiotic inspection method (SIM; de Souza, Leitão, Prates, & da Silva, 2006; de Souza, Leitão, Prates, Bim, & da Silva, to appear) and the communicability evaluation method (CEM; Prates, de Souza, & Barbosa, 2000; Prates, Barbosa, & de Souza, 2000; de Souza, 2005)—are remarkably different. One of the main differences is precisely their wide spectrum of analysis, leading evaluators to consider *the big picture*. Another is that SIM and CEM are semiotic engineering methods (de Souza, 2005), having inherited from this semiotic theory of human–computer interaction (HCI) an emphasis on communication and signification processes rather than on cognitive ones. In particular, semiotic engineering brings HCI designers onto the stage of HCI, since the *system*'s messages in user–system communication are actually *meant* and *expressed* by those who conceived the system.

In order to give our readers a preview of this shift in focus and perspective, which will be extensively explained and discussed in this book, we begin by contrasting the gist of semiotic engineering with that of cognitive engineering (Norman, 1986). Because the former provides the theoretic foundation for SIM and CEM and the latter much of the foundation for user-centered design (UCD) and usability evaluation, the contrast should help readers anticipate the share of contribution and innovation that semiotic engineering methods can be expected to bring about.

Cognitive engineering defines HCI as a goal-oriented traversal of two *gulfs*—the execution gulf and the evaluation gulf. The whole process is defined in seven iterative stages (Norman, 1986, p. 41): establishing the overall goal for interaction, forming immediate intentions for using the system, specifying the action sequence (or plan) to achieve intentions, executing actions, perceiving the system state following execution, interpreting perceived states, and evaluating the interpreted state in view of the overall goal and current intention.

The striking feature of this perspective compared to prior "user input, system output" traditional models of user–system interaction is that *all* stages of the process defined by cognitive engineering involve actions performed *by the user*. The system performs only implicit and subsidiary actions. The agent of all seven actions defined for traversing the execution and evaluation gulfs is *the user*. This model thus perfectly expresses the radical commitment with users that lies at the heart of the UCD perspective.

Another important feature of cognitive engineering is the centrality of rational goal-related mental activity in five out of seven stages of the process. The cognitive processes and workload necessary to achieve each stage provide various parameters and targets for *usable* systems. HCI designers must therefore build systems that facilitate the cognitive processes and alleviate the mental workload required to achieve the supported tasks. The end-product of the cognitive engineering of HCI is *the system image*, where various types of controls and displays are combined to promote the smooth traversals of the two gulfs. In Norman's (1986, p. 47) words:

> In many ways, the primary task of the designer is to construct an appropriate system image, realizing that everything the user interacts with helps to form that image: the physical knobs, dials, keyboards, and displays, and the documentation, including instruction manuals, help facilities, text input and output, and error messages.

Notice, however, that cognitive engineering theory does not account for what *designers are doing*, why, and how. Researchers and professionals can only use the theory to verify if or explain why certain design choices lead to "an appropriate system image" (Norman, 1986, p. 47). The theory itself cannot directly describe and explain the process of generating such choices, because the cognitive engineering model of HCI is only about the *users'* actions—not the designers'. It contributes to HCI design by helping researchers and professionals gather a body of experimental knowledge about design choices that have been proved to be *appropriate* or *inappropriate* in various types of circumstances. These take the form of design principles, guidelines, and heuristics, and usability evaluation is the method that has contributed the most to building this knowledge base.

Building *appropriate* system images can, however, turn into an overwhelming task. Consider, for example, the following interactive scenario. Fabio has recently acquired his first laptop personal computer (PC). The preinstalled operating system in it is Windows XP and one of the features Fabio is eager to test is downloading music from the Internet to his own PC library. Fabio has seen his colleagues at work do it in their computers, and he has even downloaded music to a friend's PC, which he recently borrowed for 2 weeks. He thus knows that he must use Windows Media Player (WMP) for the task. Thus, he launches WMP, and as soon as he connects to the Internet, he finds out that he must make decisions about the downloading process. The WMP interface at this stage is shown in Figure 1.1.

Notice that the "Options" dialog in Figure 1.1 is mainly guided by natural language text, implying that there is some conversation going on involving "you" (the system) and "I" (the user). This perception comes from text like: "[You, the system] Download usage rights automatically when I [the user] play or sync a file" (see third option in the "Enhanced Playback and Device Experience"). However, the roles of "you" and "I" in this conversation become considerably ambiguous. For example, although "I" play or sync a file (i.e., I, the user, execute the action), it is apparently "you," the system, the one that does everything else: displays media, updates music, downloads usage rights, checks the need to refresh files, sets the clock, sends player ID, and saves file and uniform resource locator (URL) history. The only other exception is that "I [may] want to make Microsoft software and services even better" (see Customer Experience Improvement Program option).

Moreover, all the statements about "you" and "I" are pre-fixed by a checkbox—an interface control by means of which "the user" commands the system to do something in some particular way (e.g., "[You] send unique player ID to content providers"). This is part of the WMP interface language *grammar*, which users should acquire with as little difficulty as possible. However, what is the checkbox before "I want to help make Microsoft software and services even better" commanding *the system* to do? Is this a *grammatical* expression in the interface language? Why?

Continuing with our scenario, let us assume that Fabio decides to inspect his privacy settings further, and following the instructions regarding "Enhanced Content Provider Services," he decides

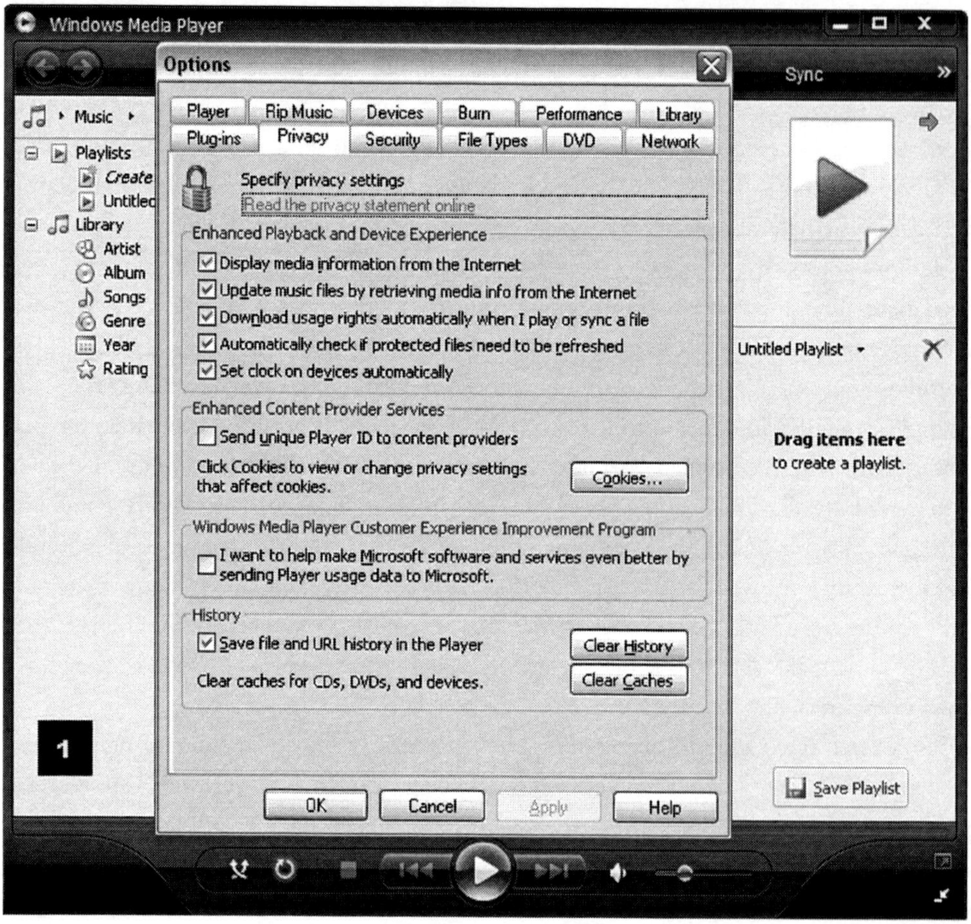

FIGURE 1.1: Windows Media Player (Windows Media Player 11, 2006 Microsoft Corporation, all rights reserved) privacy and security options for downloading.

to "click Cookies to view or change privacy settings that affect cookies." When he clicks on the Cookies button, he gets the message shown in Figure 1.2.

Now *somebody* is talking to Fabio, saying this: "You are about to change privacy settings." Apparently Fabio is no longer talking to WMP, because the message he gets says: "The Player uses Internet Explorer to communicate connection and logging information." WMP would not refer to *itself* as "The Player." Who is talking to Fabio, then? Also, notice that WMP had said that by clicking on "Cookies," Fabio could *view* or *change* privacy settings. Now, it seems that anything Fabio does will necessarily change and affect not only WMP, but other systems as well.

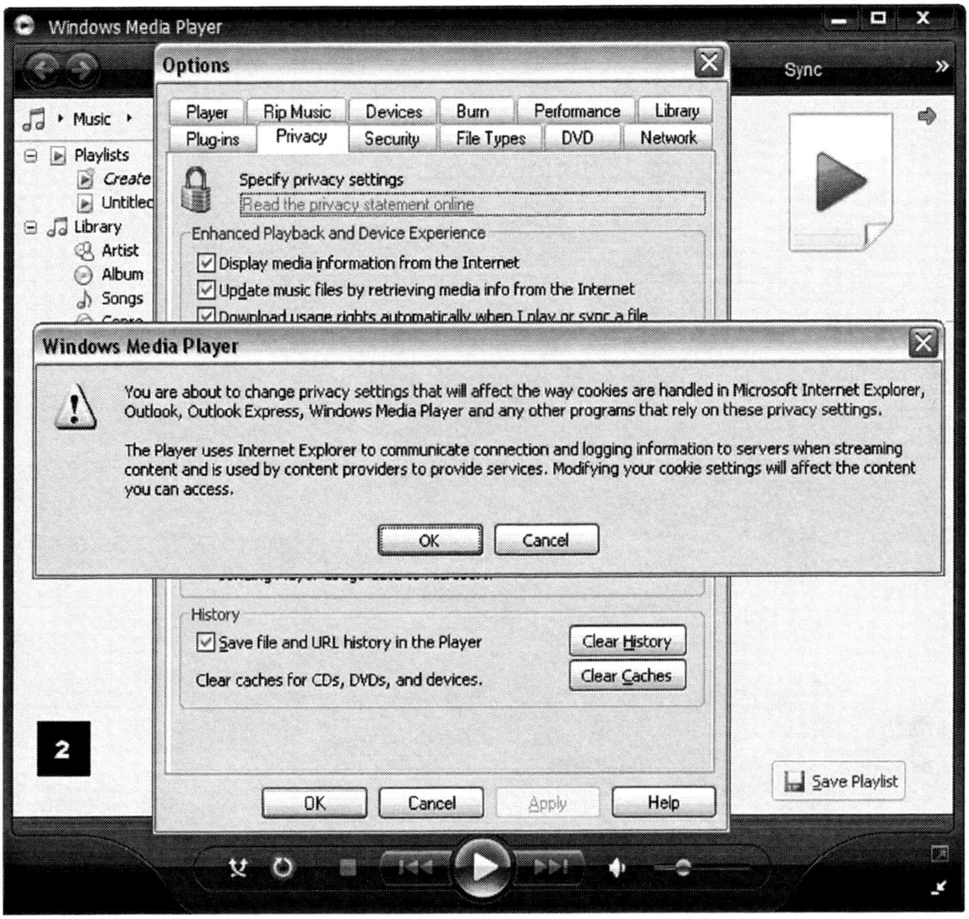

FIGURE 1.2: A system message sent to Windows Media Player users.

If Fabio dares to proceed despite all threats, he will land on an "Internet Properties" dialog (see Figure 1.3) with a lot of technical jargon in it (e.g., Internet zone, compact privacy policy, third-party cookies) and no "Help" button. This intimidating dialog will most probably deter Fabio from doing anything other than "getting out of this conversation" immediately, and probably forever.

What aspects of how cognitive engineering theory informs HCI design can be illustrated with this microscopic WMP interface issue? It is, of course, extremely difficult (if at all possible) to translate all the intricacies and dependencies of operating system parameters into *an appropriate system image*. The concepts that a user needs to understand in order to choose between different options have no parallel in their daily experience, no analog, no easy metaphor. Additionally, users

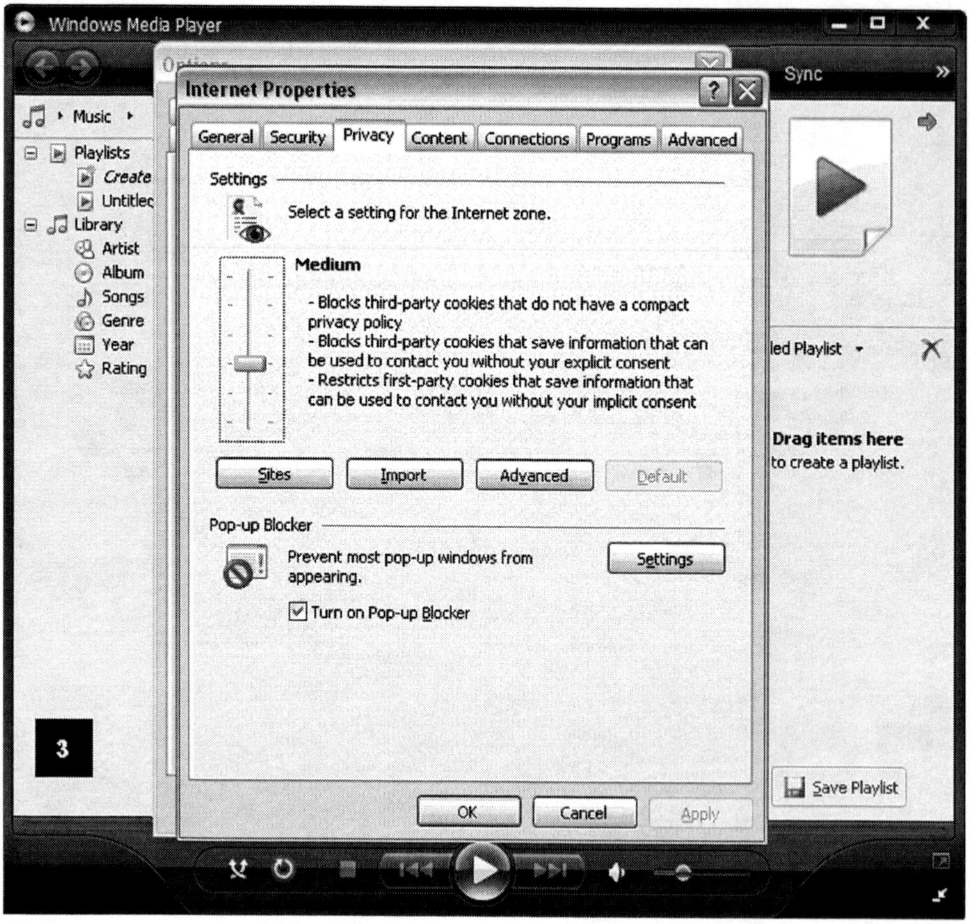

FIGURE 1.3: Internet Properties (Microsoft Windows XP Professional, 2002 Microsoft Corporation, all rights reserved) dialog encountered while interacting with various Windows applications.

typically *do not want* to know about these things, which clearly are blunders in their way to doing what *they want* (like surfing the Internet, downloading media files, chatting with friends on the other side of the world, buying various things from online stores, and so on). Thus, in order to build an appropriate system image, designers have to face the challenge of finding a conceptual model that will support users in doing what they *do not* want to do in the first place.

Similar situations are abundant in human daily life: we often have to *persuade* people to do things that they do not want to do and help them along the way. One of the most commonly used strategies is to *talk them into doing it*, that is, starting a conversation with the aim of showing them the advantages and/or need to take action. However, because the cognitive engineering model of HCI does not map onto *a conversation* (since there is only one speaker in sight—*the user*), designers do not really get much guidance from this theory when building this portion of the WMP interface. When they need to integrate, in the system image, the *images of others*, who have specific goals and intentions regarding the user (e.g., Microsoft's analysis of usage data to improve software and services, a third-party's initiative to contact the user without his or her explicit consent, etc.), cognitive engineering must pass the floor to other theories with more to say about this issue.

Semiotic engineering (de Souza, 2005) is one such theory. It embraces the challenge and proposes a different model of HCI, compared both to cognitive engineering and UCD models, on the one hand, and early user–system dialog models, on the other. In semiotic engineering, HCI is defined as a process of "communication about communication." In it, *system designers are telling users about how and why to interact with the system they have built*, as well as about other aspects of their design vision. All is communicated through the system's interface, which thus voices the designers' messages to users as interaction unfolds. The system represents the designers at interaction time and thus achieves designer–user communication.

The designers' communication must not be verbose; neither must it necessarily adopt a tutorial tone. In Figure 1.3, for instance, a simple dialog box, with its tabs, labels, and controls, is *communicating* to users that the designers' view of Internet activity involves: security, privacy, content, connections, programs, as well as general and advanced issues. If the user is a novice, for example, the designers are communicating that he or she might consider asking for help to understand and make decisions about "advanced" topics. Also, by providing default values, designers are implicitly recommending a fairly "safe" configuration for Internet usage.

Figure 1.3 also shows that *the system* mediates the designers' communication with users. The interface *voices* the designers' messages and serves as their *proxy* at interaction time. Therefore, designers, system, and users are all in communication, a process achieved by means of various types of computer interface signs. The fundamental activities accounted for by this model of HCI are sign production and sign interpretation in computer-mediated communication, which explains the semiotic nature of our theory.

With respect to the illustrative scenario used above, semiotic engineering sets out to inform designers about how to elaborate efficient and effective communication with users, so that they will have a clear notion of what is happening during interaction. Of particular importance is the fact that, unlike in natural face-to-face conversations, computer-mediated conversations of the sort illustrated above require that HCI designers have a very broad view of the conversational context and that they anticipate all conversational turns that are necessary for smooth interaction. This anticipation and subsequent encoding of solutions and possibilities in various sorts of computational interface signs is what HCI design is about, in our view. For example, designers should provide the communicative means necessary for interaction if the user wants to know *what happens* when a unique player ID is (or is not) sent to content providers.

In conclusion to this brief contrast between the cognitive engineering and the semiotic engineering views of HCI, we can summarize the main differences with two points. By adopting the cognitive perspective, designers are likely to have trouble with the enormous complexity of finding an appropriate system image that will support all the *user* steps in deciding how to configure privacy parameters for downloading media content from the Internet. As noted above, the user is not even interested in doing the task. The cognitive investment to understand the threats and intricacies of online security options is something most users would rather *not do* and *do not know how to do*. Many designers resolve this problem by allowing users not to bother with such details and making decisions for the users. This is typically followed by intimidating warnings against the risks of changing such decisions, as seen in the WMP example. The downside of this choice is that it alienates users from an important part of contemporary computer literacy—being able to understand and create protection against network security breaches.

By adopting the semiotic perspective, however, designers gain knowledge about their active role in interaction in order to help users understand issues and decide what they want or need to do. They also gain knowledge about the kinds of communicative strategies they can use, as well as about the kinds of signs that can be used in computer interfaces, the limits of computational sign interpretation and generation, and ultimately about the constraints and conditions of this particular kind of computer-mediated human communication. To give the reader a flavor of the possibilities emerging from a semiotic perspective, we propose a redesign of the Privacy tab in WMP Options dialog (see Figure 1.4). In it, the designer expresses his anticipation of critical conversations that users might wish to have about data coming in and going out of the user's machine and provides interface elements to trigger such conversations in an easy way.

Designers express their message through interface signs like words, icons, graphical layout, sounds, buttons, links, and drop-down lists. Users discover and interpret this message as they interact with the system. Interface signs, along with their dynamic behavior, are the sole means available for designers to get the user to understand what the software does and how to use it. Semiotic engineering proposes a *structure* for the designers' global message to users. Through interaction, it must communicate, implicitly or explicitly, the following contents:

1. Who are the system's users? (i.e., for whom the system has been designed);
2. Which user needs, expectations, preferences, and motives the designers have taken into consideration? (i.e., why the system has been designed in this or that way);
3. What is the system, how does it work, and why? What goals and effects are compatible with the designers' vision of the system they have built? (i.e., what range of intent and activities does the system support, in which contexts, and how can they be achieved?).

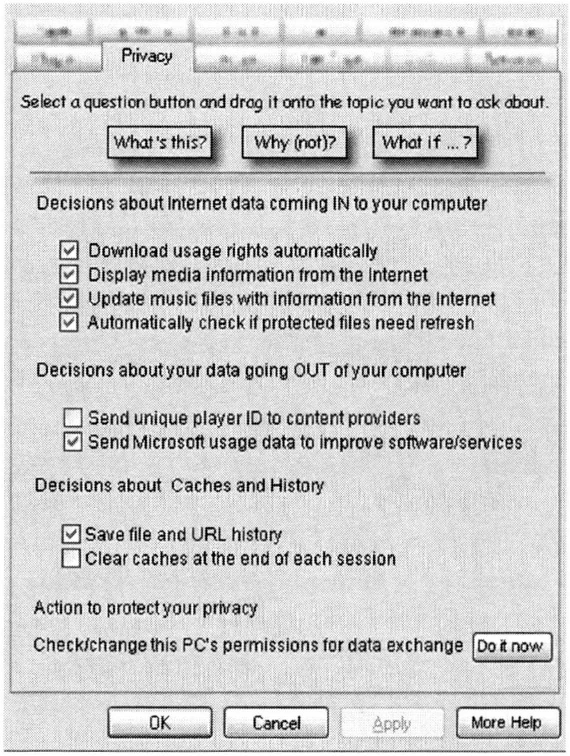

FIGURE 1.4: An illustrative redesign of communication about Privacy options.

Moreover, as suggested on p. 3, even if systems' interfaces share many interactive patterns (e.g., many systems use hyperlinks to communicate that there is more information available about the terms appearing in the link), every system has a unique interactive language whose semantics is determined by the system's unique semantic model. This language, which must be learned by users in a process that resembles second-language acquisition, is also delivered through the interface as an important part of the designer's[1] message to users. Thus, an important part of item (3) above is the communication of the very code in which the designers' global message to users has been signified. This is the language that must be employed by users to communicate back with the system. Figure 1.4 gives us a glimpse on how this communication is achieved. Once designers *tell* users that they can select and drag the question they want to ask onto the topic about which they want to learn more, they are teaching users new interactive vocabulary and grammar that, supposedly, is

[1]When using the expression *the designer*, we are referring to an entire *design team*. Alternate singular and plural forms are used freely throughout this book to emphasize the fact that both individual designers and teams have to produce a *unified* message, which will be communicated to users at interaction time.

also used elsewhere in the system. Hence, users acquire new linguistic knowledge and can thus use it to communicate back with the system. This is ultimately the reason why, in semiotic engineering, HCI is defined as *an interactive and progressive communication process about how to communicate with the system, when, why, and to what effects*. In this process, designers, system and users are equally involved and share the same role—they are all communicating parties (technically referred to as *interlocutors*).

Going back to where we started, to the recent calls for tools and methods that promote and support innovation in HCI, the contrast between cognitive engineering and semiotic engineering shows at least three significant points about our theory. First, with its focus on communication and signification, and with a model of HCI that explicitly assigns an active role to designers at interaction time, semiotic engineering is *an innovative theory* of HCI. Second, the content of communication from designers to users, as mentioned above, transcends topics related to users' goals, plans, and actions, especially because designers can now talk about *what they have done and why*, that is, about their design vision and the values in it. This sort of communication was identified by Erickson (1995), who discussed it with respect to the design process and showed how stories could be used to achieve *design evangelism* in companies and organizations. Semiotic engineering is proposing that these *stories* be told also (and mainly) to users, during interaction, by means of specific interactive strategies and sign systems. This is a particularly important feature when introducing innovative technologies, the theme discussed by Greenberg and Buxton (2008). Finally, in methodological terms, usability evaluation methods like heuristic evaluation (Nielsen, 1994) and even exploratory methods like the cognitive walkthrough (Wharton, Rieman, Lewis, & Polson, 1994), for example, are not appropriate to analyze the complexity of communication and signification processes taking place in HCI. On the one hand, their focus is typically placed on *cognitive* issues rather than on *communicative* ones. On the other, because the HCI theories and approaches that provide the foundations for these methods do not assign a role to designers during interaction, they are by definition not equipped to analyze the adequacy and consistency of the designers' communication to users through interactive discourse expressed by various kinds of static and dynamic interface signs.

In his discussion of innovation and creativity support tools, Shneiderman (2007) explicitly mentions that methods like ethnography, for example, can give researchers and designers a broader perspective on technology, and thus promote valuable insights. Thus, we now can ask if ethnography is a good method to evaluate the communication and signification processes that constitute the focus of semiotic engineering. Cooper, Hine, Rachel, and Woolgar (1995) say that "a prolonged period of intense immersion in the culture best enables the ethnographer to experience the world of her subjects, and hence to grasp the significance of their language and actions for (. . .) the production and consumption of technical artifacts" (p. 12). As this passage shows, there is no doubt that ethnography can be used to analyze aspects of HCI that are centrally important for semiotic engineering. For example, it can help us find out how users *repurpose* technology designed for different purposes and ends, like using Microsoft Excel (a spreadsheet application) as a rudimentary database system. What ethnography misses, however, precisely because this method typically operates from the bottom up, with no a priori interpretive models and schemas, is the persistent focus on semiotic structures and processes in which designers, systems, and users are involved and which constitutes the unit of analysis of semiotic engineering. The ethnographer is *not* committed to observing, collecting, and interpreting data in accordance to preexisting theoretical concepts; rather, theoretical formulations are *derived* from ethnographic research. Focusing solely and specifically on a priori issues provided by a specific theory of HCI can make more directed and agile contributions to the process of design, as we intend to show in the following chapters.

Semiotic engineering has created its own methods to evaluate HCI. As mentioned at the beginning of this introduction, SIM and CEM have been specifically designed to evaluate designer–user communication in HCI. Both methods can be used in technical and scientific contexts. In technical contexts, they can be used to improve the quality of designer–user communication of specific systems. The purpose, focus, and other circumstances of evaluation are dictated by professional needs as well as industrial or commercial interests. In scientific contexts, however, the purpose, focus, and circumstances of evaluation are dictated by research questions and methodological soundness. The immediate purpose of using the methods is mainly to advance knowledge.

This book presents a thorough description of each method followed by an extensive case study. The study shows, in great depth and detail, the kinds of results that each method can produce, both when used in isolation and when combined with each other. Although, as the title of the book suggests, there is an emphasis on the *scientific* use of SIM and CEM, HCI professionals with an inclination to reflect about their work and develop their own knowledge about the field may certainly benefit from the reading.

In Chapter 2, we present a historical perspective on semiotic engineering and define the main concepts in the theory. Then, in Chapter 3, we introduce SIM and CEM, describing the main steps in each method and illustrating them with brief examples. In Chapter 4, we present a case study, carried out in 2008, with a digital audio editor called *Audacity*.[2] In Chapter 5, we discuss the lessons learned when using SIM and CEM to evaluate interaction. Finally, in Chapter 6, we briefly express our view of where semiotic engineering is heading in the near future.

* * * *

[2] Audacity software, 1999–2008 Audacity Team. The name Audacity is a registered trademark of Dominic Mazzoni. Web site: http://audacity.sourceforge.net.

CHAPTER 2
Essence of Semiotic Engineering

This chapter begins with a brief historical perspective, which provides relevant information about the intellectual traditions from which semiotic engineering sprang. Then, we present the main concepts of the theory, which will be necessary for understanding the definition, illustration, application, and discussion of semiotic engineering methods.

2.1 BRIEF HISTORICAL PERSPECTIVE

Semiotic engineering was first proposed in the early 1990s (de Souza, 1993) as an *approach* to designing user interface languages. Influenced by existing semiotic approaches to HCI at that time (Andersen, Holmqvist, & Jensen, 1993; Andersen, 1997; Kammersgaard, 1988; Nadin, 1988), as well as by Winograd and Flores's (1986) language–action perspective (LAP), it viewed HCI as metacommunication—a process where HCI designers were sending a one-shot message to users about how and why to communicate with the system in order to achieve a certain range of goals and effects. "Metacommunication" and "one-shot message to users" were the two terms giving semiotic engineering a unique identity compared to existing work at the time. Both the semiotically oriented community and the LAP community were talking about *computer-mediated human communication*, in general. Semiotic engineering, however, stressed the fact that in HCI, some singularities were in place. Computer-mediated communication was *about communication* between users and system, its possibilities, constraints, and effects. It was clearly a case of metacommunication. Moreover, semiotic engineering gave prime importance to the fact that in this context, the directionality of the process was different from other computer-mediated communicative processes such as communication in online communities or organizational workflow systems. When communicating with users, designers encode their message entirely into a computer system; the message is gradually unfolded by users in interactive contexts, but the users never actually *talk back to the designers* over the same channel of communication. Hence, the designers' message as defined originally was a "one-shot message," and the consequences of this particular feature were far from trivial in the context of human communication.

The theoretical foundation of semiotic engineering at the time was almost exclusively Eco's *Theory of Semiotics* (Eco, 1976), with special emphasis on the parameters proposed in his theory of sign production. Very briefly, Eco characterized semiotics as the *logic of culture* and defined two fundamental processes to be accounted for: signification and communication. Signification, in his view, is the product of systematic and culturally motivated associations between contents and expressions. Communication is the process through which communicating agents explore the possibilities of various signification systems in order to generate expressions that are meant to achieve a very wide range of goals and effects. A fundamental feature of communication is that expressions *do not have to* (although they *may* and usually *do*) conform to content-form associations established by existing signification systems. Jokes, puns, and figurative speech are good examples of this phenomenon.[1] The production of signs in communication, Eco argued, could be characterized by a small set of parameters, whose values helped distinguish a very large range of communicative settings.

So in 1993, the *semiotic engineering of user interface languages* was described as a semiotic conceptual framework "within which many design issues can be explained" (de Souza, 1993, p. 753). The framework was built around Eco's parameters for sign production, and the concept of metacommunication and the idea that designers were actually communicating with users at interaction time were the seed of a full-fledged semiotic theory of HCI. It was published in a book 12 years later (de Souza, 2005).

The early years of semiotic engineering involved various kinds of efforts to translate the idea of designer-to-user metacommunication into conceptual tools that could be used in HCI *design*. Following the concepts proposed in 1993, a number of semiotically inspired models were proposed to aid in the design process of different kinds of interactive systems and application domains [e.g., groupware (Prates, 1998), end-user programming (Barbosa & de Souza, 2001; Cunha, 2001; da Silva, 2001), online help systems (Silveira, 2002), generic desktop applications (Leite, 1998)]. In 2000, we proposed a method for HCI *evaluation*, the communicability evaluation method (Prates, de Souza, & Barbosa, 2000; Prates, Barbosa, & de Souza, 2000; de Souza, Prates, & Carey, 2000), and in subsequent years, a considerable part of our research aimed at tracing metacommunication in interactive computer systems and developing methods and frameworks to help in its design and evaluation (Barbosa, Leitão, & de Souza, 2005; Barbosa, 2006; de Souza, Nicolaci-da-Costa, da Silva, & Prates, 2004; de Souza et al., 2006; Leitão, de Souza, & Barbosa, 2007).

Over the years, semiotic engineering became clearly different from other *semiotic approaches* to HCI (e.g., Andersen, 1997; Baranauskas, Salles, & Liu, 2003; Kammersgaard, 1988; Nadin, 1988). This difference also includes Jorna and van Wezel's approach by the same name (Jorna, 1990, 1994). René Jorna and we independently developed two different kinds of "semiotic engineering."

[1] For illustration, consider the meanings of the expressions 'inverse' and 'poured' in the following sentences: "a backwards poet writes inverse"; and "the man shamelessly poured his lies into the audience's ears."

Jorna's was originally closer to artificial intelligence (AI) and cognitive science. His ideas were later developed and applied to decision support systems and organizational semiotics (Jorna & van Wezel, 1995; van Heusden & Jorna, 2001) rather than HCI.

Semiotic engineering also became different from our own original version of the concept (de Souza, 1993). We stopped using semiotic theories to analyze HCIs, and semiotic engineering became *itself* a theory of HCI. In other words, instead of applying existing semiotic ontologies and methods to investigate how humans interact with computer artifacts, we developed our *own* ontology and methods and defined a distinctive unit of investigation.

Of course, the theory is built on semiotic foundations and is strongly influenced by pragmatics (Searle, 1979; Leech, 1983; Peirce, 1992–1998). However, it is incommensurate with the foundational theories in important ways. Whereas the latter can be used to investigate many other objects—like *media*, culture, language use, to name only a few—semiotic engineering can only be used to investigate the nature, the structure, the processes, and the effects of *designer-to-user metacommunication* in the context of interaction between people and computer-based technologies.

In 2005 the first complete account of semiotic engineering was published internationally in the form of a book (de Souza, 2005). There was a tangible concern to prove the theory's value in the context of professional HCI design tasks. The culture of HCI at the time—as is probably the case to date—was dominated by practical industrial demands for producing more usable systems for improving the users' experience. As a consequence, most HCI research was driven by the need to provide HCI designers with the appropriate *tools* to make the right decisions. A vast number of design guidelines, patterns, frameworks, models, techniques, and variations thereof, as well as numerous ways to evaluate the results of using such design tools, occupied most of the research landscape in the field. As a consequence, the first full-fledged presentation of semiotic engineering covered a very large scope of issues and addressed a wide community of *professional practice*, that of "reflective practitioners." Treading on Don Schön's path (1983), in 2005, semiotic engineering was trying to help HCI professionals internalize certain interpretive and analytical practices. Semiotic engineering proposed a set of *epistemic tools* that should not be used to give directly answers to design problems, but to increase the problem solver's understanding of problems and alternative solutions.

Since then, we have been continually working to make practical contributions to HCI designers and developers from an epistemic perspective (e.g., de Souza & Cypher, 2008). This tendency, however, somehow overshadowed the relevance, the implications, and the direct contribution of the theory to another community of practice, that of HCI researchers and students. So in this book, we present a detailed description, illustration, application, and discussion of methods specifically designed to evaluate metacommunication in HCI. By so doing, we fill an important methodological gap found in previous presentations of the theory.

2.2 METACOMMUNICATION PERSPECTIVE

As illustrated in Chapter 1, semiotic engineering is considerably different from many cognitively inspired theories, which focus on explicit or implicit learning and reasoning processes. It frames the object of investigation as a matter of *communication*, rather than *learning*, and thus places *designers* at the center of the process. This should not be taken to mean that users are any less important than designers. Improving the users' lives and meeting their needs and expectations constitute the ultimate reason for all the design and development enterprise. However, a theory that describes a process where designers play a first-person role is likely to give them different tools and insights about what they are doing and about the resources and opportunities to improve it, as compared to theories that describe processes where they play a third-person role (users being the exclusive first-person players). In other words, most HCI theories focus on what happens on the users' side. Semiotic engineering focuses very specifically on how designers communicate the end-product of what has happened on their side because of their knowledge and expectations regarding the users. This should be enough to show that this theory does not *compete* with user-centered theories, but articulates important aspects of design and use contexts within the same perspective (Norman, 2007).

This integrative perspective is the consequence of viewing HCI as an instance of metacommunication. A system's interface is actually telling users an important message about how they can or should use the system, why, and to what effects. The essential content of the message can be paraphrased by a generic template—called *the metacommunication template*—that says:

> "Here is my understanding of who you are, what I've learned you want or need to do, in which preferred ways, and why. This is the system that I have therefore designed for you, and this is the way you can or should use it in order to fulfill a range of purposes that fall within this vision."

The metacommunication template sums up what designers are communicating to the users through systems interfaces. The first person "I" refers collectively to *the designer* (or, in most cases, the *design team*), and the second person "you" refers collectively to *the user* (or, more appropriately, the *user population*). There are several important points to explain and justify with respect to the essence of the metacommunication message above. The first is that in communication, the role of the receiver is *as important as* that of the sender. In ideal communication, the sender actively produces signs to express his communicative intent to a particular addressee, the receiver, who by capturing the sender's message behaves in such way that she achieves the sender's intent. Ideal communication, however, is closer to the exception than the rule. Human experience is heaving with communication that breaks down halfway through, because senders do not get their message across to the receivers, receivers assume that senders are saying something and they are not, and no matter how much they try, senders and receivers do not understand each other, etc. Therefore, for metacommunication to be *in place* (even if in far from ideal conditions), it suffices that one of two things happen when users interact with computer systems: either that designers mean to tell *something* to users (i.e., to get users to behave in a particular way as a result of being exposed to intentionally produced signs); or that users take a particular course of action because they believe they are *being told* something that justifies their behavior. Our first point is then that metacommunication happens regardless of the designers' or the users' degree of awareness that they are actually communicating with each other through computer systems interfaces.

The second very important point that we must emphasize is that, of course, designers are not physically present at interaction time. Therefore, *they* are not delivering their message to the users. The *system* is. Hence, the system represents them at interaction time. It is, in semiotic engineering terminology, *the designers' deputy*. All that the system communicates must have been planned for at design time and implemented in the form of a computer program in subsequent development stages. Thus, the system's *interactive discourse* is a computational version of the conversations that designers would have with users in order to achieve the ultimate communicative intent in design, namely, that users understand, enjoy, and benefit from the design product. The content covered by such conversations is summarized in the metacommunication template.

The third point that we must clarify is that communication is achieved by a very wide range of signs. According to Peirce (1992–1998), a sign is anything that somebody takes *to stand for* (hence to "represent" and to "mean") something else. Peircean semiotics has two important elements that we should emphasize: one is that taking something to stand for something else is a process influenced by previous knowledge, intuition, and even instinct (Santaella, 2004) and the other is that the result of this semiotic process is open to correction in the presence of counterfactual evidence. The latter is intrinsically related to a particular type of reasoning procedure, which lies at the basis of sense-making activities and Peirce called it *abductive reasoning* (or *abduction*, for short). The openness of signs—always exposed to subsequent corrections and adjustments, minor or major—sustains a whole theory in which meaning is not a value, but a continuous process of interpretation also known as *semiosis* (Eco, 1976; Peirce, 1992–1998). According to this theory, all previous exposures to a particular sign that we are about to interpret—as well as our exposures to signs that we associate to it by habit, circumstance, or intuition—influence our interpretation. They add novel, even if subtle, meaning elements to our perception and understanding of the sign in question. Therefore, both designers and users continually mean subtly different things by the interface signs that they use in metacommunication. For example, over a long period of design and redesign, a designer's interpretation of a menu entry like "export file" can take many different meanings (e.g., there may be new file formats to export to, new parameters controlling the quality of conversion, etc.). Likewise, over a long period of use, a user's interpretation of a "Cancel" button can take many different meanings (e.g., do not effect actions signified on this window, interrupt this conversation, close this dialog box, etc.). However, computer interpretations of interface signs, unlike the designers' and users', are causally determined by a computer program, where a priori procedures and functions inexorably produce the same types (and very probably also the same tokens) of meaning elements. So, for example, a computer program is prepared to interpret an unlimited set of concatenated characters as file names. No matter how many times the program executes the interpreting functions; it will always produce the same data structure associating representations that correspond to file contents with representations that correspond to memory addresses. Computer meanings are always fixed, whereas human meanings always evolve. Consequently, the designer's deputy at interaction time is incapable of reproducing the semiotic processes that enable and legitimate the human communication that it mediates. In other words, metacommunication between designers and users is subject to computational constraints that semiotic engineering is prepared to describe.

Finally, although the most obvious evidence of metacommunication message elements that designers send to users through computer systems interfaces is associated to verbal signs (button labels, menu entries, tool tips, etc.), nonverbal signs can also be very efficiently used in metacommunication. The most popular example in HCI is the desktop interface, where visual representations and cursor-controlled operations on them *communicate* and effect many basic commands in a file management system.

The theoretical elaboration of metacommunication processes is not an isolated fact in the HCI landscape. We have already mentioned how LAP and previous semiotic approaches to HCI have influenced semiotic engineering. But there are more research pieces composing the mosaic of evidence that HCI can be consistently viewed as designer-to-user metacommunication. One is the work by Rheinfrank and Evenson (1996), who discuss the importance of *design languages*. According to the authors, "natural languages are used to generate expressions that communicate ideas; design languages are used to design objects that express what the objects are, what they do, how they are to be used, and how they contribute to experience" (p. 68). Another important related work is that by Nass and coauthors. In *The Media Equation*, Reeves and Nass (1996) provide extensive evidence that people respond to computer screens in the same way as they respond to human communication in natural social contexts. Sundar and Nass (2000) also report interesting empirical research results showing that even savvy computer users assign anthropomorphic qualities to user interface messages. Finally, we should mention the work by Fogg (2003), who proposes that computers are "persuasive technology" and can be used to change and affect people's behavior. Putting all of these pieces together, we see that the theoretical account of metacommunication proposed by semiotic engineering actually resonates with a number of other research perspectives and results that are not necessarily founded in semiotic theories (or at least do not explicitly acknowledge the fact).

The object of scientific investigation in semiotic engineering is more precisely defined as the set of *all* the computer-encoded conversations that the designer's deputy can have with users at interaction time, and *only* those. In other words, designer–user conversations at design time, although critically important for the success of computer systems, *do not* constitute per se an object of investigation for semiotic engineering. They are aptly investigated by other theories and approaches such as activity theory (Nardi, 1996), contextual inquiry (Beyer & Holtzblatt, 1998), and participatory design (Müller & Kuhn, 1993), among others. Likewise, computer-encoded conversations that are not consciously meant by the designers, but can nevertheless be held by the designer's deputy and users at interaction time, do constitute a legitimate object of investigation for semiotic engineering.

There are three distinctive classes of signs in the designer's deputy's interactive discourse: static signs, dynamic signs, and metalinguistic signs. Static signs are interface signs whose meaning is interpreted independently of temporal and causal relations. In other words, the context of interpretation is limited to the elements that are present on the interface at a single moment in time. For example, layout structure is a static sign and so are menu options and toolbar buttons.

Dynamic signs are bound to temporal and causal aspects of the interface, namely, to interaction itself. They emerge with interaction and must be interpreted with reference to it. For example, when a user selects the option "save as . . ." of a menu "file," systems typically exhibit a dialog window with a conversation about the file's name, location, format, etc. The causal association between the menu selection and the dialog that follows it is a dynamic sign, one that can only be expressed over time.

Static and dynamic signs are intrinsically related. Static signs stimulate the user to engage in interaction with the artifact; they help the user anticipate what the interaction will be like and what consequences it should bring about. Dynamic signs confirm or disconfirm the user's anticipation. The meaning of static and dynamic signs is explicitly informed, illustrated, or explained by signs of another class—metalinguistic signs. They refer to other interface signs, static, dynamic, or even metalinguistic (in recursive reference). Typically, they come in the form of help or error messages, warnings, clarification dialogs, tips, and the like. With metalinguistic signs, designers explicitly communicate to users the meanings encoded in the system and how they can be used.

Bearing in mind that the object of investigation for semiotic engineering is the whole spectrum of the designer's deputy's interactive discourse at interaction time, designers should pay close attention to how they integrate metalinguistic signs into the system's interface. For example, when help communication includes hyperlinks to the developers' Web site, the user's interpretation of metalinguistic signs found on the Web site will naturally be referenced to his or her ongoing interaction with the system, which brought about the need or opportunity for further clarification. But because Web site materials and system development are often carried out fairly independently by separate groups of people, it is not unusual to detect communicative breakdowns when the user asks for help during interaction. Hence, the design of metalinguistic signs is of prime interest in semiotic engineering investigations.

In 2005, when semiotic engineering was first presented as a theory of HCI (de Souza, 2005), we introduced and discussed the theory's ontology, epistemology, and methodology. Here, we will only summarize the main points of the original presentation.

There are four general categories in the semiotic engineering ontology, which are signification processes, communication processes, interlocutors involved in signification and communication, and the design space, wherefrom HCI takes its shape. Signification processes involve signs and semiosis (defined on p. 17), both collective and individual. Communication processes will involve intent, content, and expression, as well as two differentiated levels of realization: direct user–system communication and mediated designer-to-user metacommunication. The interlocutors involved in HCI are thus designers, systems (the designers' deputies at interaction time), and users. Finally, the design space is characterized in terms of senders, receivers, contexts, codes, channels, and messages.

As the ontological categories above suggest, semiotic engineering is a communication-centered theory of HCI. Because communication involves speakers and listeners and message senders and receivers and because designers, systems, and users participate in metacommunication, semiotic engineering bridges two important ontological *divides* that other HCI theories explicitly or implicitly establish. One is the divide between designers and users, and the other is that between users and systems. All three are *interlocutors* in the two levels of communication achieved at interaction time. This is a unique feature of semiotic engineering, which, rather than shifting the center of attention from users to designers as it has sometimes been mistakenly said to propose, integrates into a single theoretical framework concepts and phenomena that are treated in isolation by different theories.

By adopting many definitions springing from Peirce's (1992–1998) and Eco's (1976) semiotics, semiotic engineering inherits many epistemological commitments of these foundational theories. The most important one for the purposes of this book is a split perspective on meanings exchanged in metacommunication. When produced and interpreted by humans, meanings are, as seen on p. 17, open and subject to constant evolution and unlimited semiosis (Peirce, 1992–1998; Santaella, 2004). However, when generated and interpreted by systems as a result of symbol manipulations effected by computer programs, they are algorithmically specified and thus established and delimited a priori. This duality of perspectives has at least two important consequences.

The first consequence is that, along the process of metacommunication, human signification is *narrowed down* to computational symbol processing. While reproducing the designer's interactive discourse, the designer's deputy can only effect mechanical manipulations of semiotic materials. Therefore, fundamentally important elements of human signification involving interface "signs" (e.g., the evolving meaning of an iconic representation associated to a toolbar button) cannot be computed. For example, even if the designers' and the users' signification of "" have so much in common that they can be virtually said to *be the same*, the "system" does not have the least notion of what minimize, maximize, or close window mean to the designers or the users. Consequently, although designers and users may agree on how "" *should* behave in exceptional situations like the abnormal termination of a process activated on the window controlled by "," all that the designer's deputy can do is to trigger the execution of functions associated to each of the three buttons, using the same contextual parameters as it always does. In this sense, computer mediation in metacommunication can *introduce* breakdowns that designers and users would not experience in nonmediated communication with one another.

The second consequence is that, although human meanings *cannot* be strictly predicted or fully inspected because they are constantly evolving (even if in minute details), computer meanings *can*. Hence, there is at least one aspect of signs involved in metacommunication that is open to extensive scientific investigation, because it is fixed and fully specified prior to its actualization during interaction. This is the designer's deputy's interactive discourse, whose origin and destination is bound to *human* signification and communication processes. Hence, the investigation of metacommunication discourse must not lose sight of this duality, which brings about important methodological challenges that will be discussed in the next chapters.

CHAPTER 3

Semiotic Engineering Methods

SIM (de Souza et al., 2006; de Souza et al., to appear) and CEM (Prates, de Souza, & Barbosa, 2000; Prates, Barbosa, & de Souza, 2000; de Souza, 2005; Sharp, Rogers, & Preece, 2007) have been originally proposed as applications of semiotic engineering theory to support professional HCI activities. As already mentioned, inspired by Schön (1983), de Souza (2005) defined SIM and CEM as *epistemic tools* meant to help professionals in developing reflective, interpretive, and analytical HCI design practices.

Over the years, SIM and CEM have been used in academic projects more often than in professional practice. Their repetitive use in research (e.g., de Souza, Laffon, & Leitão, 2008; Leitão, de Souza, & Barbosa, 2007) eventually showed that both methods have an important role to play in the development of the very theory from which they have sprung. Beyond strictly technical contributions, they allow investigators to expand their knowledge about HCI, in general, and semiotic engineering, in particular, helping them to reformulate concepts, identify gaps, and propose new items for an ongoing research agenda. The possibility of using SIM and CEM as scientific investigation procedures has turned into an attractive item, because this would provide a closure to the iterative cycle of research, which begins with the acquisition of knowledge by means of systematic and rigorous procedures (scientific methodology), continues with the conceptual elaboration of acquired knowledge (scientific theory), and is completed when knowledge is used to achieve a wide range of social purposes (scientific knowledge application and development of technical knowledge). Therefore, results of the scientific application of SIM and CEM should expand both scientific and technical HCI knowledge.

Scientific methods used in HCI research should contribute to one or more of the following goals:

- A new account of known problems, using theoretical concepts that support the formulation of relevant research questions;
- The identification of new solutions, partial or complete, generic or specific, to known problems and challenges;

- The identification of new problems and challenges; or
- The formulation of new theories, concepts, models, or methods.

Different theories and approaches are likely to require different scientific methods that are fully consistent with their ontological and epistemological commitments. Because SIM and CEM are based on and circumscribed by the ontology of semiotic engineering, they are meant to support scientific investigations exclusively centered on designer-to-user *metacommunication*. In other words, they can only help us gain knowledge about the designers' communications computationally encoded in the form of interactive messages presented in a computer system's interface. SIM and CEM support the exploration of the designer's deputy's interactive discourse by analyzing and reconstructing metacommunication and by evaluating the *communicability* of such discourse.

Defined in 2000 as "the distinctive quality of interactive computer-based systems that communicate efficiently and effectively to users their underlying design intent and interactive principles" (Prates, de Souza, & Barbosa, 2000, p. 32), the concept of communicability has been subtly changed and complemented as semiotic engineering gradually evolved into a full-fledged theory of HCI.

The original definition of communicability quoted in the paragraph above mainly referred to operational conditions. At the time, we were mainly trying to detect evidence of communication of design intent and design principles in interactive computational artifacts. However, the interlocutors involved in the process were not explicit in that definition.

In 2005, we discussed the metonymical formulation of the concept defined above. The definition transferred "to the design (the product) a capacity that is in fact expected from the designer (the producer)" (de Souza, 2005, p. 113). In the discussion of the metonymy embedded in the previous definition, we proposed that:

> Communicability can (. . .) be more technically defined as the designer's deputy capacity to achieve full metacommunication, conveying to users the gist of the original designer's message. (. . .) Communicability applies to both interpretive and expressive codes that the designer's deputy handles for generating and interpreting messages during situated interaction with users. (de Souza, 2005, p. 114)

This *new* definition connects communicability more clearly to the ontology of semiotic engineering. By stressing that the communicative process in focus is that between users and the designers' deputy, we reiterate the notion that, in this theory, designers and users share the same ontological status—they are interlocutors at interaction time. This change in theoretical formulation brings about an important consequence for semiotic engineering methods. By clearly defining the roles of sender and receiver of metacommunication (the designers as *senders*, represented at interaction time by the designers' deputy, and the users as *receivers*), the new definition allowed us to elucidate that each method explores a different angle of metacommunication (de Souza et al., 2006). SIM explores the *emission* of metacommunication, seeking to reconstruct its content, expressions, and targeted receivers. CEM, in its turn, explores the *reception* of metacommunication, seeking to identify, by means of user observation, empirical evidence of the effects of the designers' messages as they are encountered at interaction time.

More recently (de Souza et al., to appear), the definition of communicability has been refined once again. Different applications of SIM and CEM (de Souza et al., 2008; Leitão et al., 2007), followed by methodological studies about qualitative and quantitative research paradigms, pointed to the impending risks of suggesting, even if implicitly, that semiotic engineering methods can (or should be expected to) *measure* efficiency and efficacy in metacommunication. Both terms are commonly used in the context of quantitative research, and without further clarification, there might be important misunderstandings about communicability and the methods used to evaluate it. So in order to avoid misunderstandings, we added a very straightforward definition of what we mean by efficient and effective communication, saying that we refer to communication that is organized and resourceful (*efficient*) and achieves the desired result (*effective*).

We can thus summarize the definitions of SIM and CEM by saying that both are scientific methods that can be used to explore the communicability of the designers' deputy interactive discourse. And they do so by supporting the analysis and reconstruction of metacommunication. They allow researchers to identify computationally encoded strategies with which designers communicate design intent and design principles and to verify if communication is organized and resourceful (de Souza et al., to appear). Additionally, researchers can verify the effects of metacommunication and decide if they are consistent with evidence of what designers mean to say by their designs. The metacommunication template (p. 16) that paraphrases what designers are telling the users through systems interfaces helps researchers in their exploration.

Finally, knowledge generated by SIM and CEM is validated with the same procedures used in qualitative research (Cresswell, 2009; Denzin & Lincoln, 2000; Patton, 2001). Validation is achieved by triangulating results, a process in which replicability and reliability criteria are replaced by plausibility and consistency among results produced with different means and resources. At this point, consistency among methods is critically important. SIM and CEM can only be consistently triangulated with nonpredictive theories and research results because semiotic engineering is itself incompatible with the notion of *predicting* human meanings and interpretations, the origin and destination of metacommunication.

After this preliminary methodological discussion regarding the nature and application conditions of SIM and CEM, we begin to describe the procedural steps of each method. We will provide only brief illustrations of each step because Chapter 4 presents an extensive and complete application of SIM and CEM in a case study, where abundant details are provided.

3.1 SEMIOTIC INSPECTION METHOD

SIM is an inspection method conceived to explore the designer's deputy's interactive discourse with an emphasis on its *emission*. It aims to reconstruct metacommunication using the metacommunication message template as a guide (p. 16). Regarding a particular research question under investigation, SIM allows us to reconstruct the designer's message in its entirety. This is possible precisely because it is an *inspection* method, and the researcher is fully in charge of carefully selecting and examining all significant interactions, as well as of analyzing and interpreting them. The researcher can then explore all the communicative potential of interactions, including the identification of design intent, communication contents, expressive choices, and alternative paths, both successful and unsuccessful.

Guided by the metacommunication template and following rigorously all the method's steps, the researcher delimits his field of analysis and engages in an interpretive process leading to the reconstruction of the designer's message. He must examine important aspects of the emission of metacommunication, summarized by the following orientation questions:

- What is the designer communicating?
- To whom is the designer's message addressed?
- What effect(s) does the designer expect his communication to cause?
- How is the designer signifying his communication?
- What expectations does the designer have about what users will want to communicate to the system?
- How, where, when, and why does the designer expect users to engage in communication with the system?

SIM application begins with a preparation phase and is then conducted in five core steps, shown in Figure 3.1, from bottom up:

1. The analysis of metalinguistic signs;
2. The analysis of static signs;
3. The analysis of dynamic signs;
4. A comparison of the designer's metacommunication message generated in the previous steps; and
5. A final evaluation of the inspected system's communicability.

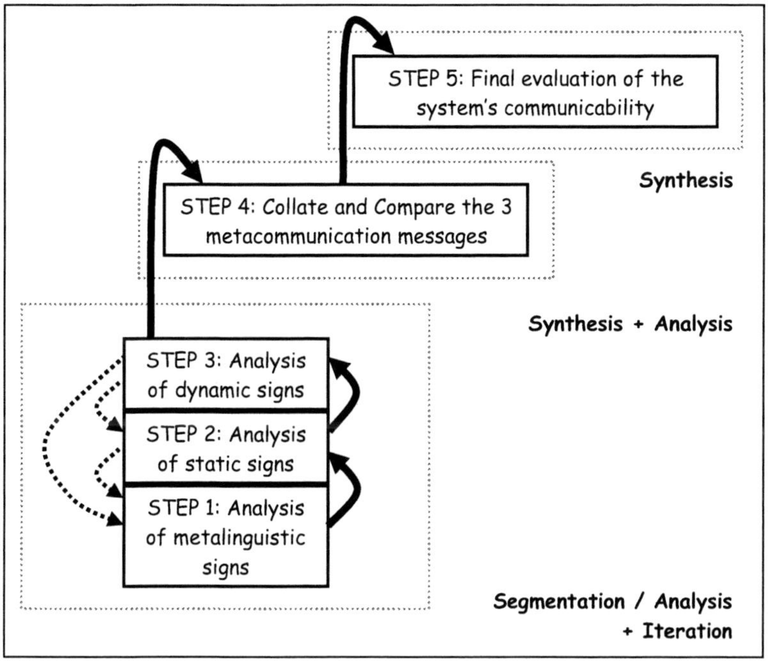

FIGURE 3.1: Five core steps of SIM.

In steps 1, 2, and 3, which are iteratively applied, the researcher does a segmented analysis of the system, one for each of the three classes of signs that are specific to semiotic engineering: metalinguistic, static, and dynamic signs (see p. 19). This segmented analysis actually *deconstructs* the metacommunication message, allowing the researcher to inspect in great detail what and how the designer communicates with each type of sign. In steps 4 and 5, the researcher engages in the activity of *reconstructing* the metacommunication message by comparing, integrating, and interpreting the data collected in previous steps of the method. At this point, the researcher can fill out the metacommunication template and is able to articulate his empirically based conclusions with the research question that calls for investigation.

Like other qualitative methods used in scientific research, the method requires an important step in the end of the whole process, namely, the triangulation of results. As discussed in the beginning of this chapter, methods and theories used in this final stage must be carefully selected, so as to avoid methodological inconsistencies in the overall investigation process.

We now present a description of SIM steps, with brief illustrations of how they are achieved in practice.

3.1.1 Preparation

The initial step is to decide whether the research question can indeed be explored with SIM. The method deals with open questions related to communicative strategies that are evidenced in designer–user metacommunication. For example, one such question might be: "Which communicative strategies are used to orient users when they have to do complex spatial operations on visual objects?" Notice that this one is an *open question*, which SIM can help answer. However, this method is not particularly productive when dealing with closed questions like: "which communicative strategy is best when users have to do complex spatial operations on visual objects: metaphors to express operations or interface wizards to guide interaction?"

Once the researcher has decided that SIM is an adequate method to explore his research question, he must choose the computational artifact(s) that will provide the empirical base for semiotic inspection. For example, when investigating communicative strategies used to support complex spatial operations, he can choose Microsoft Word[1] (MS Word), which allows users to do complex page, text, and graphics manipulations while editing and formatting a digital document.

SIM preparation also requires that the researcher carries out an informal inspection of the chosen artifact, aiming to establish the *focus of analysis*. SIM, as other qualitative methods, privileges an in-depth analysis. Hence, establishing the appropriate focus is critical, and because the process under investigation is *communication*, the researcher must necessarily establish minimal contextual conditions for his analysis. He must identify: (i) who are the intended users of the system, and (ii) what are the top-level goals and activities that the system supports. Without knowing at least (i) and (ii), he would not be able to begin to analyze metacommunication.

The last step in the preparation phase is to elaborate the inspection scenario (Carroll, 2000), which projects the research question upon the territory of possible interactions with the chosen artifact(s). This step fully contextualizes the researcher's analysis and operationalizes the research. Following the example with MS Word, we might elaborate an inspection scenario where the user is preparing a printable version of a document with pages edited in "portrait" and "landscape" orientation. An important element of the scenario, which causes complex spatial operations to take place, is that the user wants to number pages sequentially and have all page numbers appear in the same position and orientation in the printed document. In Figure 3.2, we see the illustration of a successful interaction. This particular scenario presents a wealth of issues to analyze because it is the result of a long chain of interactions involving not only the configuration of page orientation (portrait and landscape) but also the appropriate control of sequential page numbering with necessary reformatting of page numbers in "landscape" orientation. In Figure 3.3, we sketch the spatial

[1]Microsoft Office Word 2003. ©1983–2003 Microsoft Corporation. All rights reserved.

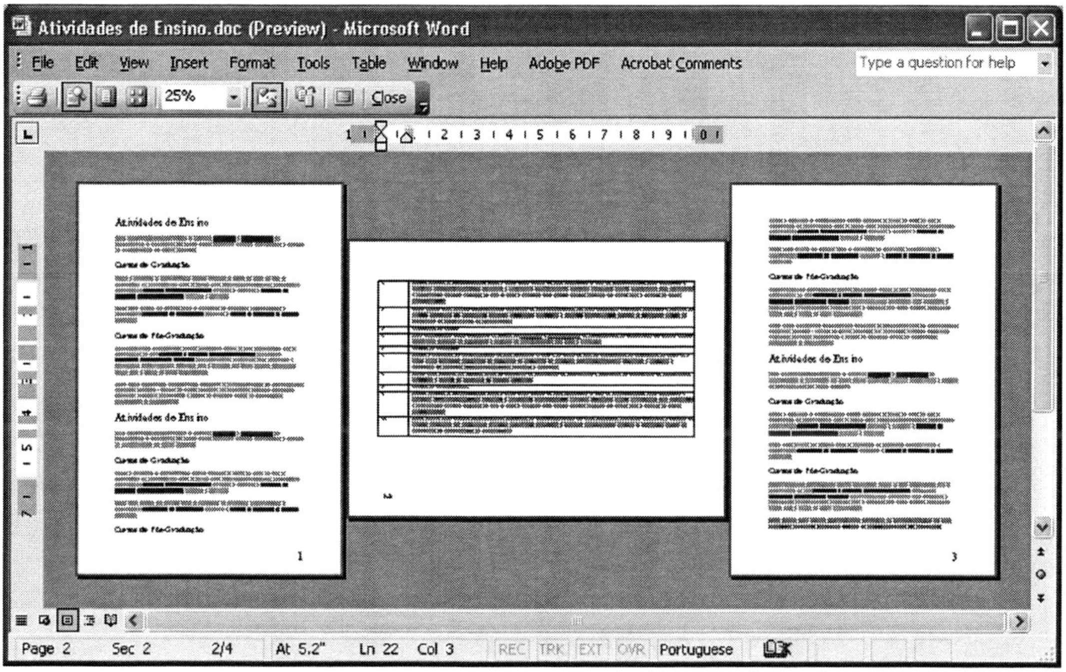

FIGURE 3.2: An illustration of a successful page-numbering strategy to print documents with pages in portrait and landscape orientation with MS Word. Microsoft Office Word 2003. ©1983–2003 Microsoft Corporation. All rights reserved.

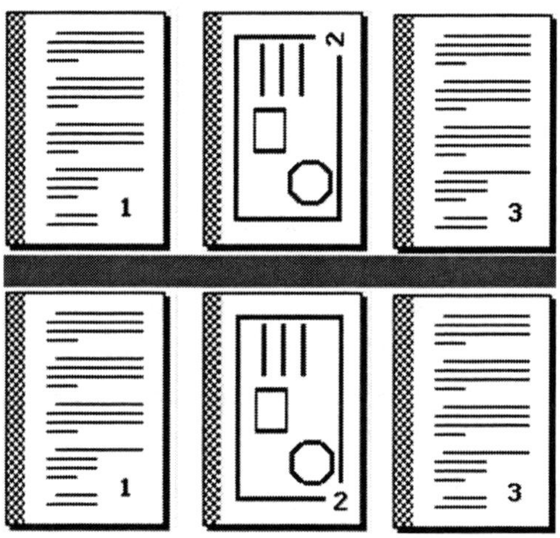

FIGURE 3.3: A sketch of page-numbering manipulations for correct printing.

configuration resulting from "new page orientation" alone (*top*) and from "new page orientation + new page-numbering specifications" (*bottom*).

3.1.2 Analysis of Metalinguistic Signs

A small illustration of what is inspected in this step is to type "Page setup" in MS Word help request text box. As a result, the researcher retrieves various fragments of the designer's discourse *about* MS Word, that is, various fragments of metacommunication where metalinguistic signs communicate the designer's perspective on this topic. In one of them, following links in the offline help material, the designer tells users that they can: select the pages that they want to change, open the "File" menu, click on "Page Setup," open the "Margins" tab, click on "Portrait" or "Landscape," and in the "Apply to" box, choose "Selected text." Part of the action involved in the inspection scenario is explicitly explained. However, the designer's message does not allude to the effects of page reorientation on the layout of printed pages. As the researcher proceeds with his exploration of help material, he finds communication about where to position numbers on the page ("top," "bottom," "center," "left," "right"), but there is no explicit communication about how to format page numbers so that they look as shown in Figure 3.2. In order to get this information, the researcher must realize that he has to switch topics of conversation and frame the interactive problem as one of "text direction," not "page setup." So one of the results of the analysis of metalinguistic signs in this illustration is that conversations about "page setup" do not communicate all of the critical aspects regarding the final layout of a printed page, even if the missing item in the MS Word example is, for all practical purposes, the orientation of the "page number" (not of generic "text").

3.1.3 Analysis of Static Signs

In this step, the researcher revisits the inspection scenario, now looking at metacommunication achieved by static signs alone. These are mainly expressed by screen layout, menu structures and options, images, text, dialog boxes, etc. It is important to analyze these elements *statically*, that is, separate from temporal and causal relations that constitute the object of analysis of another step in this method. The researcher must have in mind the communicative context defined by the inspection scenario (i.e., the designer's interlocutor, the interlocutor's goals while communicating with the system, the semiotic universe that this interlocutor is expected to have access to, etc.). Again, at the end of this step, the researcher fills out the metacommunication template and registers what designers are telling users by means of this specific class of signs.

In the MS Word example that we are using to illustrate the main aspects of an inspection with SIM, the researcher will examine metacommunication looking at how it is achieved by static signs appearing on this system's interface. He will inspect different menus such as "File" (to change page orientation), "Insert" (to add page numbers at the desired position), and "Format" (to change the direction of text in the box where page numbers are shown). In each case, the researcher will go down two or three levels of menu selections, and in his inspection of the "Format" menu, he will find that users have to take new action. As shown in Figure 3.3, the default page-numbering format (*top*) introduces an error when a document mixes portrait and landscape orientation for pages: the page number appears on top of the *rotated* page, instead of at the bottom, with different orientation than the other printed page numbers. The message sent through metalinguistic signs in the online help, which was analyzed in the previous stage, tells users how to change text orientation from "ᘔ" to "2." However, as the ongoing inspection will show, the user needs to capture and integrate communication about an additional topic, not clearly related to page setup or text orientation, the "binding" of footer and header contents between pages. If page footers are *bound* with previous and next, the specific changes in page-numbering layout that will make the printed landscape page look *right* will affect all other *bound* pages and make them look *wrong*.

This is yet another piece of communication, not clearly expressed by metalinguistic signs related to page numbering and page setup, and expressed by static signs that do not clearly refer to each other. Note that the "topic" of communication is page numbering and page setup, but the result that the user wants to achieve in the inspection scenario is related to the binding of footers and headers across pages. In Figure 3.4, we see the static signs that communicate the designer's message about how to complete the page number formatting process. In the "Headers and Footers" toolbar, the  sign should tell users that they *can* release the binding between footers and headers of subsequent pages and, in this particular scenario, that this is *necessary* when they print documents with pages in different orientations but want headers and footers to look the same on all printed pages.

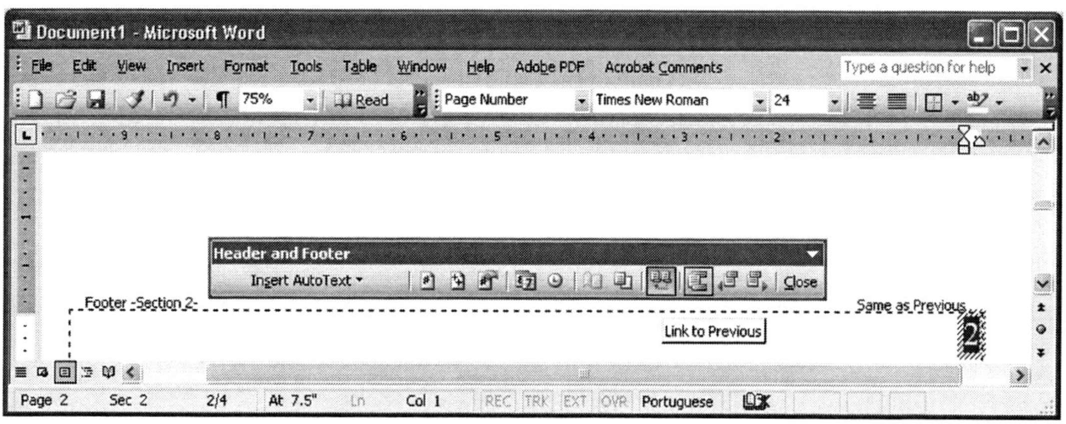

FIGURE 3.4: MS Word toolbar communicating the designer's message about the need and opportunity to bind or release headers and footers on subsequent pages.

Although this is only a small example of what is involved in the analysis of static signs, it illustrates how a relatively frequent problem in metacommunication can be identified in this step. Static signs expressing the designer's messages about tightly related topics are dispersed throughout the interface, without explicit (or sufficiently strong) reference to each other. Notice that although we can easily capture in a single image the "problem" and the "solution" under inspection (see Figure 3.3), the corresponding pieces of metacommunication in MS Word are dispersed through various menus and direct manipulation contexts, not always linked to each other (if linked at all). As will be seen in the next step of SIM, capturing the designer's message is only possible with the support of dynamic signs.

3.1.4 Analysis of Dynamic Signs

In this step, the researcher revisits the system once again, now examining *interaction*, which enables communication through dynamic signs. He will analyze signs that express transitions between system states, animations, and further signification of system behavior over time. As in previous steps, the result of his analysis is registered in the metacommunication template, filled out with messages expressed only by dynamic signs.

In the MS Word example, the researcher will see that the communication strategy known as "What you see is what you get" (WYSIWYG) relies heavily on dynamic signs to give users instant feedback. For example, if he selects the text box where the page number is shown and drags it across the page to another position, the contiguity between his ongoing action and successive system states is a dynamic sign that tells him (or *means*) that in order to reposition the object on page, he must only move it to the desired location. Once the user gets this message, he can engage in interactive strategies, such as *successive refinements* or *trial and error*, in order to get the entire metacommunication message about how to achieve the effect sketched in Figure 3.3. The researcher will then realize that metacommunication through dynamic signs helps the user integrate dispersed metacommunication achieved through static signs, which naturally leads him to the fourth core step in SIM.

3.1.5 Comparison of Segmented Metacommunication Messages

In step 4, the researcher will finally collate and compare the results of segmented metacommunication analysis. The aim is to detect inconsistencies and/or consistent relationships and patterns between elements collected in segmented analysis. As already suggested, in the small example we are using to illustrate SIM steps, the researcher will see that the metacommunication template is filled out differently in each step. Messages communicated through metalinguistic, static, and dynamic signs do not completely coincide. For example, help messages do not refer to all significant items that direct manipulation and WYSIWYG feedback show to be related. Likewise, static signs in menu structures and dialog boxes do not prompt the user to make certain kinds of associations required for a successful run of the inspection scenario.

3.1.6 Final Evaluation of System's Communicability

In core step 5, the researcher finally evaluates the communicability of the system, by reconstructing a unified metacommunication message and judging the costs and benefits of communicative strategies identified in previous steps. In conclusion to our MS Word example, the researcher will detect communicability problems with certain spatial operations that are of great relevance in the domain of electronic publishing and text editing. More specifically, he will conclude that the designer's strategy to get his message across by stimulating users to engage in *trial and error* or *successive refinements* with WYSIWYG and direct manipulation techniques may come at the expense of systematic long-term learning. Users may begin to experience fortuitous successes, without really understanding the interactive principles that lead to desired results. This is of course a problem in metacommunication, especially if users misinterpret certain side effects of operations as being intended design principles. From this small example, the researcher will also realize that metacommunication in MS Word does not explore the possibility of creating redundancies within the same sign class or across different sign classes, which could stimulate the users' semiosis in productive directions. For instance, within the segment of metalinguistic signs, when querying offline help for "page setup" in MS Word, the user is invited to select one of three alternative subtopics: "Select paper source," "Select paper size," and "Change page margins." If the user follows the second or third alternative, she will find links to "sections" and "section breaks." However, in none of the three options does the designer refer to page numbers or textual material in headers and footers, which can be so deeply affected by changes in page orientation and margins. These are treated as separate, unrelated topics. And across different segments, like static and dynamic signs, metacommunication in MS Word is also lacking in useful redundancies. For example, page orientation dialogs offer a sketched preview (a dynamic sign) of the page when the user selects one orientation or the other. However, the sketched pages corresponding to each orientation, which are static signs per se, do not include representations of headers and footers (which would immediately call the user's attention to potential problems in print). So in sum, in this small example, the researcher would detect a communicability problem stemming from loose integration between communications achieved with different classes of signs.

From a scientific perspective, at the end of step 5, the researcher relates his research question to the empirical results achieved with SIM. In Chapter 4, we will present a detailed description of how this can be done, emphasizing the depth and scope of conclusions that a researcher may find. We should also remark that a *triangulation* is required to ensure the scientific validity of achieved results. Again, in Chapter 4, we will show how SIM results can be triangulated with CEM results and produce new valid knowledge in HCI research.

3.2 COMMUNICABILITY EVALUATION METHOD

CEM was the first method proposed by semiotic engineering to analyze metacommunication. Because it is based on the observation of how a small group of users interacts with a particular system, CEM imposes limitations on the analysis and interpretation that leads to the reconstruction of metacommunication. The partial reconstruction produced with CEM is based on empirical evidence of how the designer's messages are *received*.

The limits of metacommunication reconstruction are established by the interactive paths that observed users choose to follow during test sessions. When using SIM, the researcher is free to explore the computer artifact in as many different directions and for as long as it occurs to him. This allows him to make a very broad scanning of metacommunication and to reconstruct the designer's message in its entirety, given the research question that guides analysis and interpretation. When using CEM, however, the researcher is guided by what the users actually do in a single test session (which takes approximately 30 minutes). Moreover, his analysis can only refer to the *evidence* collected during these tests, which is, in fact, evidence of *failures* in the reception of metacommunication. In other words, CEM focuses on a particular class of observable phenomena in user–system interaction, namely, that referring to communicative *breakdowns*, which narrows the scope of analysis even further. Some examples can show why this is the case.

Let us suppose that a test participant is engaged in running the same test scenario as was used above to illustrate SIM. At the end of the test, she has succeeded in putting her document in the prescribed format. The researcher has not seen any evidence of a problem with the reception of metacommunication. Can he say, however, based on his observation, that metacommunication in MS Word is efficient and effective?

We argue that he cannot, because even if the user has not experienced any breakdowns while communicating with the designer's deputy, this does not mean that she thinks that the designer's metacommunication is clear, organized, resourceful, and useful. As researchers, we cannot know if the user finds the designer's messages consistent nor can we infer that the user actually understood the strategies used by the designer and got the metacommunication message in its entirety. At this point, it is useful to recall something that we have already mentioned when discussing SIM, the fact that users often succeed in doing what they want in a completely fortuitous way. They just *happen* to do the right thing, although they probably cannot explain why, or repeat the successful sequence of actions.

Therefore, the *absence* of observable problems in interaction does not necessarily mean that the designer's message was received correctly and completely. However, the *presence* of such observable problems does mean that metacommunication was not received as intended.

Even within the limits of this partial reconstruction of metacommunication, which is based on evidence of communicative breakdowns, CEM can produce a wealth of results. This is because real users' experiences always bring surprises to the eyes of researchers, who can then explore, analyze, and interpret aspects of metacommunication different than those that would have occurred to them when using SIM, for example. Although SIM allows researchers to cover a wider spectrum of issues, CEM provides them with the seed of unpredictability that is always present in *any* communicative setting, from face-to-face conversations to HCI. Thus, both methods can be used to complement each other, or they can be alternatively selected depending on specific purposes and conditions of investigation.

CEM supports a researcher's exploration of important aspects of metacommunication that can be summarized by the following orientation questions:

- How is the user interpreting the designer's communication?
- What does the user want to communicate and how can she do it?
- What effect does the user want her communication to produce?
- How is the user "signifying" her communication?
- How is the user communication being interpreted by the system (i.e., by the designer's deputy)?

After the preparation and application of user tests, CEM is carried out in three steps of analysis and interpretation shown in Figure 3.5, from bottom up:

1. Tagging;
2. Interpretation; and
3. Semiotic profile.

At each step, the researcher gradually achieves higher levels of abstraction in his analysis and interpretation of how metacommunication is received. In step 1, the researcher watches the recording of each user's session and identifies passages that are indicative of breakdowns in communication. Each passage is tagged with 1 of 13 specific utterances (e.g., "Help!," "What's this?") representing the researcher's interpretation of how the user's behavior relates to the context of interaction where it occurs.

After having tagged all passages where he sees evidence of a communicative breakdown, in step 2, the researcher begins to interpret the meaning of the whole set of dispersed tags. This interpretation is based on the presence or absence of each of the 13 tags, on their frequency and distribution across different contexts of interaction (and different user sessions), as well as on the

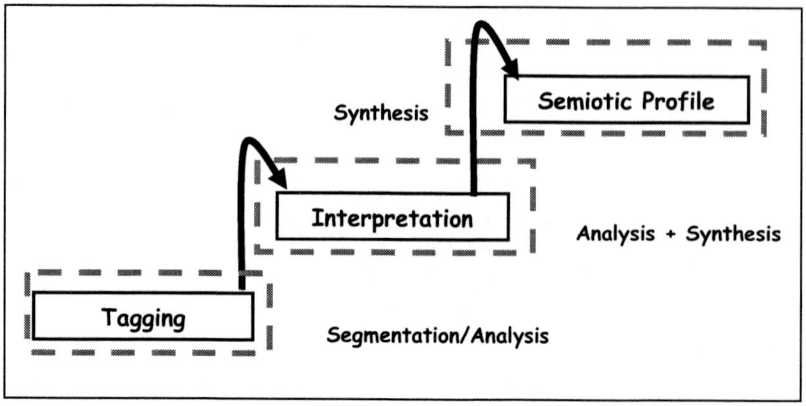

FIGURE 3.5: The three core steps of CEM.

theoretical categorization of tags according to the semiotic engineering ontology (de Souza, 2005). Then finally, in step 3, the semiotic profile step concludes the whole process with an in-depth characterization of the metacommunication message's reception.

When used in scientific research, as already mentioned, CEM results must be triangulated with other results in order to be validated.

After this brief description of the method, we now describe each one of its steps in greater detail.

3.2.1 Preparation

As is the case with SIM preparation (p. 28), the researcher must decide if his research question can be adequately explored with CEM. If so, he must choose the computational artifact(s) that will be used to collect empirical data. He must then do an informal inspection of the selected artifact(s) so as to identify the appropriate focus of investigation. It is possible to replace this informal inspection by a full application of SIM, in which case CEM can be used to triangulate results.

Because CEM involves user observation, it is also necessary to define the criteria for selecting test participants. The number of participants recruited for the test, as is usual in qualitative research that aims to do in-depth studies, is small. In our experience, the selection of 6–10 participants has been typically sufficient to reach the point of saturation required for data collection in qualitative research (Seidman, 1998).

With the results of his informal inspection and the definition of the participants' profile, the researcher proceeds to elaborate the test scenario, the script of a pretest interview (if necessary), the informed consent form, as well as the necessary hardware and software infrastructure for the test (Sharp et al., 2007). Occasionally, the researcher may wish to prepare a script for part of the posttest interview that is always carried out, but, as will be explained in the next paragraph, is contingent to his observations during the test. He then does a pilot test to verify if the test application needs further adjustments and applies the test itself.

3.2.2 Application

CEM must be applied by two evaluators. One is primarily in charge of attending the user during the test. He is there to help the user and to ensure that the test equipment is fully functional. He is also a privileged observer of the user's verbal reactions and attitude, which are important ingredients for subsequent interpretations of collected interactive evidence. The second evaluator is fully concentrated on observing the test. In most situations, he is doing this using cameras or clone monitors behind a mirror window.

Both evaluators must seek to identify immediately all signs of communicative breakdowns in the user's interaction. They must take note of these during the test and, in the posttest interview that concludes the data collection procedure in CEM, try to clarify the user's reception conditions relative to the observed breakdowns. In particular, the evaluators must try to disambiguate the meaning of certain behaviors like opening unnecessary dialogs, for instance. Why did the user do this? Did she think that this was the right action to take, or was she just curious to know what the dialog was about? Note that being able to make these distinctions is crucial for the researcher, who will otherwise be prone to serious mistakes in his interpretation.

3.2.3 Tagging

In this step, the researcher watches the recording of all test sessions and carries on a segmented analysis to identify all evidence of *communicative breakdowns*. To each one of the identified breakdowns, he associates 1 of the 13 utterances (or *tags*) proposed by semiotic engineering. Tags are ordinary natural language expressions, commonly encountered in human communication, which the user *might* plausibly utter if prompted to manifest herself verbally during the test or while watching the video with the recorded session. For this reason, tagging can be described as "*putting words in the user's mouth*, in a kind of reverse protocol analysis" (de Souza, 2005, p. 126).

The 13 communicability utterances that characterize communicative breakdowns between the user and the designer's deputy are the following:

- "I give up."
- "Looks fine to me."
- "Thanks, but no, thanks."
- "I can do otherwise."

- "Where is it?"
- "What happened?"
- "What now?"
- "Where am I?"
- "Oops!"
- "I can't do it this way."
- "What is this?"
- "Help!"
- "Why doesn't it?"

As mentioned before, they express the researcher's interpretation of how observed interaction relates to the context of metacommunication reception. They have been described and discussed in greater detail elsewhere (Prates, de Souza, & Barbosa, 2000; Prates, Barbosa, & de Souza, 2000; de Souza et al., 2000; de Souza, 2005; Sharp et al. 2007). Here, we present a brief description of each, which suffices to explain what the tagging step involves.

"I give up." This utterance is used to tag interaction where the user explicitly admits her inability to achieve her goal. The general symptom of this breakdown is that the user interrupts her activity without having accomplished all of the proposed task(s). This may occur at any time during the test and is always associated to other breakdowns in the reception of metacommunication. For illustration's sake, keeping with the MS Word example used to illustrate SIM, the user might interrupt the test because she could not locate the interface control to reposition the page number on a page with landscape orientation. This would happen only after various attempts to succeed, which typically amounts to a long chain of trial and error (with evidence of other kinds of reception failures, characterized below).

"Looks fine to me." This tag is applied when the user is convinced that she has achieved her goal but, in fact, has not. The symptom of this breakdown is that the user terminates the test falling short of achieving all the tasks described in the test scenario. When asked if all tasks have been achieved, the user will say that they have. In the MS Word example, the user might successfully reposition the page number on the landscape page without realizing that this has a negative effect on the position of page numbers everywhere else in the document.

"Thanks, but no, thanks." This utterance is used when the user is aware of the designer's deputy's metacommunication regarding the types of conversations that are expected to lead to a particular effect, but chooses to do something different than is expected. Knowing what is "expected" is the result of careful examination of explicit manifestations of the designer regarding how certain tasks and operations are achieved. This is typically included in help material. Because the user gives the researcher evidence that she knows what the designer is saying, but decides to follow a differ-

ent interactive path, she *declines* the designer's invitation to engage in that particular kind of communication. Hence, there is breakdown, even if, from a cognitive point of view, there is evidence that the user is in full control of interaction. For example, in the MS Word scenario, the user might make a quick incursion into the help material and decide to do the task by splitting the document into different *sections* and then configuring the format of each section separately. This is not what help instructions tell her to do. On the contrary, the designer's explicit message is that MS Word can create and manage section breaks automatically if the user follows the suggested path. Thus, if the user decides to do something else, we take it to mean that designer and user do not share the same perspective on efficient modes of communicating with the system in that particular context of activity.

"I can do otherwise." This tag is used when the user is not aware of the designer's deputy's metacommunication regarding the types of conversations that are expected to lead to a particular effect. She then chooses to do something different than is expected, but achieves the same effect. This situation is slightly but critically different from the previous one, where the researcher should use the "Thanks, but no, thanks." tag. The breakdown tagged with "I can do otherwise." is in some respect more severe than the previous one because now the user reveals that she *has not received* the designer's message about how the system should be used in the context where she is. In the MS Word example, this tag should be used if the user decides to *draw* text boxes and other elements to reproduce visually, on landscape pages, the appearance of headers and footers on portrait pages. This might probably include editing such as *whitening* or *erasing* undesired elements in the landscape page layout, so that they *do not show in print*.

"Where is it?" This tag is used when the user expects to see a certain sign that corresponds to a particular element of her strategy, but cannot find it among the signs expressed by the designer's deputy. The user must be convinced that the sign she is looking for is the one she needs to express her current goal (otherwise, the problem is associated to another kind of breakdown). For example, the MS Word user in our scenario might think that in order to reformat landscape pages, she needs to find a "rotate" function that she can apply to headers and footers. Hence, she will probably spend some time (maybe a long time) searching for a "rotate" sign, which she will not find in this case, leading to other breakdowns tagged as "I give up." or "I can't do it this way." (see below), for example.

"What happened?" This utterance is used to tag interaction where the user repeats an operation because she cannot see or understand the evidence of the effects caused by her actions. The typical symptom of "What happened?" is the user's repeated activation of a function whose feedback is either absent or not perceived. In the MS Word example, this situation might arise if the user clicked repeatedly on the "Link to previous" sign, without realizing that this is enabling and disabling independent configuration of headers and footers in the document (see Figure 3.4).

"What now?" This tag is used when the user is temporarily clueless about what to do next because none of the designer's deputy's signs mean anything to her. The typical symptoms of "What now?" is when the user is following a random path in interaction. No connection can be traced between one interactive step and the next. The difference between a "What now?" tag and a "Where is it?" lies in the user's knowing the content she wants to express (the case of "Where is it?") or not having any notion (the case of "What now?"). This kind of breakdown can turn into a severe case of miscommunication if, during random interaction, the user cannot find a sign that will spark interpretations that will bring her back into communication with the designer's deputy and eventually lead her out of the breakdown situation.

"Where am I?" This tag is used when the user is interpreting (and potentially using) signs that belong to the designer's deputy's vocabulary, but doing so in the wrong context of communication. The main problem in this breakdown is the signification of context, which confuses the user. In MS Word, for example, trying to edit the document in "Print Preview" mode is a relatively frequent problem, especially because certain views in "Print Layout" mode, where editing is enabled, look very much like "Print Preview" pages.

"Oops!" This tag is used when the user momentarily makes a mistake and immediately corrects it. She sees that she has made a wrong step and usually activates the "undo" function immediately. However, if the attempt to correct her mistake develops into a long search for a way to cancel the effects of a slip, then it indicates a very serious communication problem.

"I can't do it this way." This utterance is used to tag interaction where the user abandons a path of interaction (composed of many steps) because she thinks it is not leading her towards her goal. The typical symptom of an "I can't do it this way." is when the user suddenly interrupts an activity she is engaged in and takes a totally different direction. When explaining the aforementioned "Where is it?" tag, we mentioned, for illustration, that a user might be looking for a "rotate" function to resolve the formatting problems with landscape pages. After doing this for a while, she might be convinced that this is an unproductive strategy and begin to use another strategy (e.g., to insert section breaks in the document, manually, and then format each section separately). Given this example, we can notice that the "I can't do it this way." tag indicates a breakdown where the user has invested much more time and cognitive effort in doing the wrong thing than "Oops!".

"What's this?" This tag is used when the user expects to see an explanatory tip or any other cue to what a particular interface sign means. In the MS Word example, this might be the case if the user deliberately *inspected* the meaning of signs on the "Headers and Footers" toolbar, with the mouse hovering over each button until the corresponding tip was shown (see Figure 3.4).

"Help!" This tag is used when the user explicitly resorts to metalinguistic metacommunication in order to restore productive interaction. She may deliberately call a help function by pressing F1 or

read documentation material offline. Although used less frequently than one might expect, online help is certainly a privileged communicative resource for designers.

"Why doesn't it?" This utterance is used to tag interaction where the user is trying to make sense of the designer's deputy's message by repeating the steps of previous unsuccessful communication in order to find out what went wrong. She does not know how to express her intent, but suspects that the sign she is currently examining is the one to be used for achieving the intended goal. In other words, the user is using experimentation to make sense of how the system works. In the MS Word scenario, while trying to reconfigure the orientation of a single page in the whole document, the user might insist on opening the "Page Setup" option of the menu file and exploring subsequent dialogs, because she does not realize that she must switch the default selection set in the "Apply to" box from "Current Section" to "Selected Text." If she eventually makes the right selection, this means that she finally *got* the designer's message and stepped out of the communicative breakdown. However, she may also move on by adopting a radically different strategy of interaction (which the researcher will tag as "I can't do it this way.").

3.2.4 Interpretation

In this step, the researcher works with tagged material, seeking to identify the main problems with metacommunication. He analyzes and organizes (or classifies) collected evidence according to four different perspectives (de Souza, 2005), which gradually lead him to more abstract levels of interpretation of the empirical data:

- the frequency and context of occurrence of each type of tag;
- the existence of patterned sequences of tag types;
- the level of problems signaled by the occurrence of tag types and sequences; and
- the communicability issues that have caused the observed breakdowns.

Analyzing the frequency and context of tags is important to help the researcher identify recurrences of breakdowns in designer–user communication. For example, a high frequency of "What now?" tags may indicate that the observed users do not *signify* intentional elements of the task they are about to do in the same ways as the designer does. Hence, they cannot formulate their intention in terms that the designer's deputy is prepared to interpret.

The identification of patterned sequences of tag types provides solid interpretive basis for the researcher when trying to detect the origins of miscommunication. For example, the identification of a patterned sequence of "What now?" followed by "I give up." indicates the origin of severe metacommunication problems, a mismatch between the users' and the designer's signification systems when expressing task-related *intent.*

Interpreting the level of problems signaled by the occurrence of tag types and sequences involves distinctions between operational, tactical, and strategic communication problems. Operational breakdowns are typically *local* interactive problems, whereas tactical breakdowns spread over longer interactive paths, requiring not only more sophisticated semiotic resources from users (who must resignify what they mean, for instance) but also more sophisticated cognitive resources (typically invested in learning activities). Cognitive efforts, however, are not analyzed by semiotic engineering. We mention them here only to point at clear opportunities for articulating CEM results with those of cognitive methods and theories. Strategic breakdowns may be fatal for technology adoption, in that they may point at fundamental misconceptions in design about who the users are, what they want or need to do, how, and why (see the metacommunication template on p. 16).

The interpretation of communicability issues evidenced by CEM is achieved with the aid of theoretical tag categorizations (de Souza, 2005) shown in Table 3.1. By relating communicative intent, content and, expression with communicative effects, the theory defines three major categories of metacommunication failures: complete failures, partial failures, and temporary failures.

Complete failures are associated to definitive, unrecovered problems in the reception of metacommunication. The user is unable to understand the designer's message conveyed by the designer's deputy's interactive discourse. This is a deep and severe problem, of which the user may be aware (tagged by "I give up.") or not (tagged by "Looks fine to me.").

Partial failures are associated to unexpected interactive paths taken by the user. The pattern of conversation between the user and the designer's deputy may be "unexpected" in different ways, although they all lead to the achievement of the user's intent. First, the designer may explicitly communicate that some other conversational path is *expected* to be easier, more efficient, or more appropriate in some other respect than the one chosen by the user. The explicit communication is normally the object of communication achieved with metalinguistic signs in very salient contexts (e.g., the content of how-to instructions shown when the user presses F1 in MS Word). Second, the user's choice may be the result of her declining *default* interactions promoted by the designer. In many cases, there may be good reasons to suppose that *default* values express the designer's suggested path for interaction. Third, the interaction may lead to success because of *side effects* of one or more conversational paths that have nothing to do with the expected topic of conversation. This is indeed a serious problem in metacommunication. Other than these, interaction may be unexpected in various ways that are totally contingent to specific aspects of the domain of application, interface style, test situation, and so on. These must be carefully examined by the researcher, so as not to mislead him to the wrong conclusions.

Temporary failures are subcategorized into three types of failures: those related to a momentary interruption in the user's ongoing interpretation and sense-making activity; those related to

TABLE 3.1: Categorization of communicability tags.

CATEGORIZATION	DISTINCTIVE FEATURE	TAG	ILLUSTRATIVE SYMPTOMS
Complete failures			
	User is conscious of failure.	"I give up."	The user believes that she cannot achieve her goal and interrupts interaction.
	User is unconscious of failure.	"Looks fine to me."	The user believes she has achieved her goal, although she has not.
Partial failures			
	User understands the design solution.	"Thanks, but no, thanks."	The user deliberately chooses to communicate her intent with unexpected signs, although she has understood what preferential designer's solutions are promoted.
	User does not understand the design solution.	"I can do otherwise."	The user communicates her intent with unexpected signs because she cannot see or understand what the system is telling her about better solutions to achieve her goal.
Temporary failures			
1. User's sense making is temporarily halted	Because she cannot find the appropriate expression for her intended action.	"Where is it?"	The user knows what she is trying to do but cannot find an interface element that will tell the system to do it. She browses menus, opens and closes dialog boxes, etc., looking for the particular sign.

TABLE 3.1 (*Continued*)

CATEGORIZATION	DISTINCTIVE FEATURE	TAG	ILLUSTRATIVE SYMPTOMS
Temporary failures			
1. User's sense making is temporarily halted	Because she does not see or understand the designer's deputy's communication.	"What happened?"	The user does not understand the system response to what she told it to do. Often, she repeats the operation whose effect is absent or not perceived.
	Because she cannot find an appropriate strategy for interaction.	"What now?"	The user does not know what to do next. She wanders around the interface looking for clues to restore productive communication with the system. She inspects menus, dialog boxes, etc., without knowing exactly what she wants to find or do. The evaluator should confirm if the user knew what she was searching ("Where is it?"), or not ("What now?").
2. User realizes her intended interaction is wrong	Because it is uttered in the wrong context.	"Where am I?"	The user is telling things to the system that would be appropriate in another context of communication. She may try to select objects that are not active or to interact with signs that are output only.

TABLE 3.1 (*Continued*)

CATEGORIZATION	DISTINCTIVE FEATURE	TAG	ILLUSTRATIVE SYMPTOMS
Temporary failures			
2. User realizes her intended interaction is wrong	Because her expression is wrong.	"Oops!"	The user makes an instant mistake but immediately corrects it. The "Undo" operation is a typical example of this tag.
	Because a many-step conversation has not caused the desired effects.	"I can't do it this way."	The user is involved in a long sequence of operations, but suddenly realizes that this is not the right one. Thus, she abandons that sequence and tries another one. This tag involves a long sequence of actions while "Oops!" characterizes a single action.
3. User seeks to clarify the designer's deputy's intended signification	Through implicit metacommunication.	"What's this?"	The user does not understand an interface sign and looks for clarification by reading a tool tip or by examining the behavior of a sign.
	Through explicit metacommunication.	"Help!"	The user explicitly asks for help by accessing "online help," searching system documentations, or even by calling the evaluator as a "personal helper."

TABLE 3.1 (*Continued*)

CATEGORIZATION	DISTINCTIVE FEATURE	TAG	ILLUSTRATIVE SYMPTOMS
Temporary failures			
3. User seeks to clarify the designer's deputy's intended signification	Through autonomous sense making.	"Why doesn't it?"	The user insists on repeating an operation that does not produce the expected effects. She perceives that the effects are not produced, but she strongly believes that what she is doing should be the right thing to do. In fact, she does not understand why the interaction is not right.

the user's momentary inability to communicate back with the designer's deputy; and finally, those related to the user's inability to understand the designer's deputy's signification choices.

The user temporarily interrupts her interpretation of metacommunication, causing suspension of the sense-making process, in three different contexts. In the first case, the user knows what she wants to do but cannot find, among the expressive possibilities offered to her by the designer's deputy's, an appropriate choice to express her intent ("Where is it?"). In the second case, the user cannot see or understand the designer's deputy's response to her communication ("What happened?"). Finally, in the third case, the user fails to capture the cues in the designer's deputy's discourse to formulate an appropriate communicative intent and proceed with metacommunication ("What now?").

Failures related to the user's momentary inability to communicate back with the designer's deputy may occur because the user is expressing herself using the wrong signs for the current conversational context ("Where am I?"), or because she is simply using signs that mean something else. Hence, the designer's deputy is not getting her intended communication. Some breakdowns in this category are quickly detected and repaired ("Oops!"), whereas others take longer to be detected ("I can't do it this way."). The user's frustration in the latter situation is likely to be high, a problem that is central in usability studies, for instance. Here again we see how semiotic engineering studies may be articulated with HCI studies originating in other theories and traditions.

Finally, when the user herself sets out to repair metacommunication problems, trying to clarify the designer's deputy's signs and patterns of communication, she is expressing a special type of breakdown. There are also three subtypes of breakdowns to distinguish in this case. The first is the user's implicit query of the *designer's* meanings typically exposed by tool tips and other formats of just-in-time information ("What's this?"). The second is the user's explicit query of the *designer's* meanings typically exposed by online and offline documentation ("Help!"). And the third is actually a discovery strategy with which the user tries to guess the *designer's* meanings by repetitive experimentation. In this case, the user consciously repeats interaction that was previously unsuccessful, trying to find hidden meanings that will eventually clarify problematic pieces of the designer's deputy's communication ("Why doesn't it?").

The categorization of communicability problems helps the researcher make the final leap in abstraction, reaching a top-level view of metacommunication issues, which is necessary to achieve the last step of the method.

3.2.5 Semiotic Profiling

In this step, an in-depth characterization of metacommunication is achieved. The designer's message can be spelled out by the evaluator, who will be able to assume the first person in discourse and speak for the designer with the following questions.

Who do I think are the users of the product of my design? The answer to this question tells the main characteristics of the listener of the designer's metacommunication message. The answer should also tell something about matches and mismatches between the designer's intended listeners and the actual listeners.

What have I learned about these users' wants and needs? The answer to this question will help spot fine-grained mismatches between what the designer intended to say with his design and what users get from it and do with it.

Which do I think are these users' preferences with respect to their wants and needs, and why? The answer to this question tells the designer's justification for the signification systems he has used, and if the decisions he made are consistent with "the real world."

What system have I therefore designed for these users, and how can or should they use it? The answer to this question will tell mainly how well the expression and content of the designer's metacommunication is being transmitted to the user.

What is my design vision? The answer to this question will tell mainly how well the design rationale has been understood (and accepted) by the user.

By answering these questions, the researcher will have in his hands a deeply detailed characterization of how metacommunication is received in the specific context of his research. Although this is not an exhaustive description of how users interpret and use the designer's message, it nevertheless provides a wealth of insights for the exercise of synthesis and abstraction that is required in scientific investigation. At this point, the researcher must relate the results obtained with CEM with the research question he set out to investigate, producing a theoretical account of his study.

We should once again call the attention to the importance of validating CEM results with triangulation procedures. This can be done with the results of other compatible procedures, originating outside semiotic engineering, or with the results produced by SIM. In Chapter 4, the reader will be able to follow the steps of a scientific inquiry where CEM was used to validate results previously achieved with SIM in the context of a case study with a digital audio editor.

* * * *

CHAPTER 4

Case Study with Audacity

As was seen in previous chapters, the scientific application of semiotic engineering methods requires that, after having defined precisely their research question, researchers choose an adequate HCI instance that prefigures significant issues related to the matter under investigation. In this case study, our research question was: *which strategies do the designers choose in order to communicate a system's basic functions and how they can or should be used?* The interest of such research question lies in that basic functions are essential to all users of a given system, regardless of their level of expertise and of how specialized their ultimate goals may be. Therefore, the quality of interaction when using them extensively determines the perceived quality of interaction with the whole product.

The next step in this study was to select a sufficiently complex computer system, so that it would make sense to distinguish between *basic* and *nonbasic* (advanced) functions. We chose to work with Audacity, an open-source freely distributed digital audio editor, with localized interfaces for different language communities. In May 2008, it was ranked on PC World's 100 Best Products of 2008.[1] In August 2008, it won the sound editing category in *InfoWorld*'s Best of Open-Source Software (BOSSIE) Awards.[2] It is very popular among nonprofessional users, and, in its developers' opinion, the editor is "easy-to-use." The system has been designed with beginners in mind, but at the same time, it offers specialized functions for generating various types of sound effects and analyzing sound waves and patterns. Plug-ins and sophisticated functions can be used by experienced users in quasi-professional contexts, as well as by beginners who are just curious about things they can learn with Audacity. Moreover, Audacity is developed and improved continually by a team of volunteers. Most of the developers are C++ programmers although, among them, several have background in HCI. Volunteer participation in open-source development gives us the opportunity to probe more challenging issues in semiotic engineering like the impact of communication infrastructure and processes in development practices, as compared to communication infrastructure and processes enabled through the system's interface. These issues will be discussed in the conclusion of our case study.

[1] See http://www.pcworld.com/article/146161-12/the_100_best_products_of_2008.html.

[2] See http://www.infoworld.com/slideshow/2008/08/165-best_of_open_so-2.html.

We started the study with SIM, doing a semiotic inspection of Audacity. Then, in order to triangulate our results, we used CEM and carried out a communicability evaluation test with a group of six first-time users of Audacity. Our findings at this stage were contrasted with evidence collected in two separate contexts: an observation of how an experienced user performed while doing the same tasks as the first-time users we observed in our tests, and interviews with three members of the Audacity Development Team. Our conclusions at the end of this study clearly show why and how a scientific application of SIM and CEM differ substantially from a technical application. On the one hand, our findings have been insightful to Audacity's designers by providing an enriched account of their communicative strategies regarding design intent. In so doing, designers can now expand their knowledge about Audacity, analyze and interpret unexpected results, and eventually improve Audacity's interface in some specific ways. However, the main contribution of the study was to identify and formulate HCI knowledge and issues that now require further investigation or that, at a higher level of abstraction, can be tied to (and benefit from) knowledge and issues discussed in other sub-areas of computer science (see Figure 4.1).

FIGURE 4.1: SIM and CEM in scientific investigation connecting HCI with CS. All images of Audacity, including logo, screen shots, and web material, are copyrighted and used with permission from the Audacity Development Team.

This chapter is organized in three sections. The first describes how we used SIM to analyze Audacity with an emphasis on the *emission* of metacommunication. The second describes how we used CEM to analyze Audacity with an emphasis on the *reception* of metacommunication. Finally, the third section discusses the triangulation of results and presents the conclusions of our study.

4.1 A SEMIOTIC INSPECTION OF AUDACITY

The semiotic inspection of Audacity's release 1.3.5 (beta version) was carried out during the months of September and October of 2008. Web site materials and programs used in the study were the ones publicly available at the time. SIM *steps* have been described in the previous chapter. What follows is the description of SIM *results* at each step, using the metacommunication template (see p. 16) to guide our presentation.

4.1.1 Preparation for the Inspection

After visiting Audacity's Web site and exploring the system, we identified the following common characteristics among Audacity's intended users:

- They are interested in digital audio editing for essentially homemade and noncommercial productions;
- They enjoy and value learning opportunities;
- They support and use open-source freely distributed software; and
- They would consider joining a community of users and developers to report problems, find out how to solve them, help developers improve the software, and/or help other users.

Other than that, as already mentioned, Audacity's designers have both novice and advanced users in mind, although their emphasis on learning opportunities lead us to conclude that they are paying special attention to novices. It is also noteworthy that they are concerned with accessibility. One link on their Web site explicitly says "Accessibility (Audacity for the visually impaired)."

Finding out which are the basic functions of Audacity poses no difficulty. When the system is first run, the user sees the *welcome message* shown in Figure 4.2. In it, the designers clearly communicate the basic functions: play, record, edit, open or save a project, and export sound to an audio file. We find out by following the corresponding hyperlink that burning a CD must be done with other programs like iTunes or Windows Media Player.

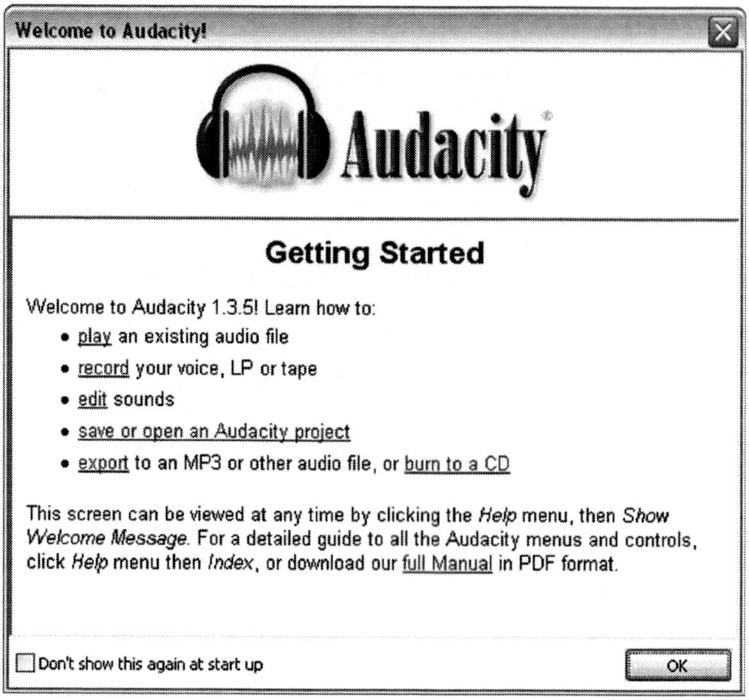

FIGURE 4.2: Audacity's welcome message.

With this information in mind, we elaborated the following inspection scenario:

Your friend Joe is a fan of ringtones. He has patiently associated different sounds to each of his 20+ cell phone contacts. The other day he was telling you how much fun he has producing his own MP3 tones for every contact using this neat digital audio editor called Audacity. It is free, he says, and easy to use. He even showed you the ringtone that plays when you call him.

You are definitely not like Joe—not as patient, not much of a cell phone enthusiast, but you love music, and you think it would be fun to dabble in digital audio editing. You have done some movie editing with family clips in the last few months, and you guess that audio editing must be about the same kind of thing.

This is a long and lazy weekend for you. So, you have decided to explore this ringtone idea. You have already downloaded Audacity and explored it quickly. You would like to send an email to Joe with a surprise package attached to it: your own homemade ringtone to replace the one he uses when you call him.

So, this is what you do:

You pick up one of your favorite MP3 files in your computer.

You use parts of it to compose a ringtone with about 1 minute of duration.

You decide to show off and add your own voice to the ringtone in two different ways:

- first, you say something like "Answer the phone, Joe" while the music is playing; and
- second, near the end of the tone, you interrupt the music, say something like "Come on, Joe! Answer the phone, will you?," and then add a last bit of the music to finish up the tone.

The scenario helped us to focus on a specific use situation, a well-defined context and activity. The tasks involved in the activity cover virtually all of Audacity's basic functions and highlight two of Audacity's targeted users' common characteristics: somebody that is interested in digital audio editing for essentially homemade productions and is willing to learn new things.

4.1.2 Analysis of Metalinguistic Signs

The analysis of online help material, including tutorials, and of explanations, instructions, and warnings provided at interaction time through dialogs, screen tips, and the like allowed us to fill up the metacommunication template in the following way.

"Here is my understanding of who you are."

Audacity has been designed for a *wide variety of users* interested in creating and editing digital audio files. Among the user profiles explicitly addressed by the development team are high school teachers and students, visually impaired individuals, podcasters, musicians, and game programmers (see Figure 4.3).

- (cc) *Audacity 1.2 Tutorials* from the official Audacity user's manual.
- (cc) *Audacity Presentation for High School Level* (PowerPoint file, also openable in *OpenOffice.org*)
  (cc) *(PDF version)*
- Using Audacity 1.3.4 with the *JAWS screen reader for Windows* - includes useful tips for the visually impaired (and others) who want to use Audacity without a mouse.
- (cc) *"Audacity 101 slides"* - PDF tutorial on all aspects of Audacity (1.2.4), including illustrations about setting up system sound on Windows and Linux.
- *How to Record and Edit Audio with Audacity* from guidesandtutorials.com. Includes information on microphones and configuring Audacity on Windows.
- *Create your own podcast* - a fine general tutorial on all stages from recording and editing to publishing on the web, again from guidesandtutorials.com
- *Introduction to Audacity* mainly for Guitarists
- *Reworking voice records*
- *How to Create BGM Loops with Audacity* - mixing two loops together in Audacity for background music in games
- *Learn Audacity in an Hour* - series of short, snappy video tutorials with downloadable AIFF files by Baynard Bailey of *Vassar College* : covers opening files, editing, using specific effects and exporting

FIGURE 4.3: Audacity's online help material for different user profiles.

In spite of such variety, metalinguistic signs appearing in the welcome message (see Figure 4.2) are clearly addressed to novice users, and to the basic tasks that they can carry out using Audacity. Given the focus of our inspection, the content of the welcome message, including text we find following the hyperlinks on it, is especially relevant. For example, as we follow the "edit" link shown in Figure 4.2, we find the following explanation:

> The main commands for editing audio are under the Edit menu (such as cut, copy, and paste) and the Effect menu (you can do things like boost the bass, change pitch or tempo, or remove noise).
>
> Audacity applies edits to selected areas of the audio track. To select a particular area, click in the track and drag the shaded area with the mouse. If no audio is selected, Audacity selects all the audio in the project window.
>
> When playing or recording, the Edit and Effect menus will appear grayed out, because a moving track can't be edited. Commands can sometimes be unavailable for other reasons too. For example, you can't run effects until you have audio on the screen, and you can't paste audio until you've cut or copied it to Audacity's clipboard. Audacity ©1999–2008

The interest of the communication above is that there are no hyperlinks in it. In other words, the message is completely self-contained, which allows us to examine the kinds of assumptions that Audacity's designers have about their audience of users. In the first paragraph, we are told that editing audio includes doing "things like boost the bass, change pitch or tempo or remove noise." The absence of further explanations, or links to them, suggests that the addressed users are assumed to know what these things are. It is however surprising to find the following passage in the same message: "For example, you cannot run effects until you have audio on the screen, and you cannot paste audio until you've cut or copied it to Audacity's clipboard." The help message above communicates that whereas Audacity users are assumed to be relatively savvy in terms of manipulating sound, they may be lacking in computer literacy. Most personal computing applications today, like text and graphics editors or digital spreadsheets, come with direct manipulation interfaces in which having a selected object prior to activating commands is always required. Hence, the last part of the explanation above is not necessary for computer-literate users.

The impression that metacommunication in Audacity is addressed to computer-illiterate users is nevertheless incompatible with the interpretation of messages in other parts of the system. For instance, in the online manual, when explaining how to select objects in Audacity, the designers say:

> In a word processor, most operations are performed by selecting a range of text (usually with the mouse), then choosing some option from a toolbar or menu, for example to cut the text,

> or make it bold. Audacity works much the same way: most operations are performed by selecting audio with the mouse and then applying some operation. Audacity ©1999–2008

Clearly, they now rely on the user's familiarity with word processing and direct manipulation. This ambivalence toward the *interlocutor*'s profile in metacommunication with Audacity can take different forms. For example, there is a special session in online documentation called "Audacity for the Impatient." This is a quick guide for people who want to jump right into the action and learn by doing. The content of this quick guide contains hyperlinks to the full manual, where more extended explanations are found. In general, hyperlinked information refers to details and advanced tasks. The effect of hypertext structure in communication may nevertheless be confusing. Readers of "Audacity for the Impatient" are supposedly *impatient*. Thus, the style of explanations and instructions on the main page and its child pages is brief, or even terse. On the main page, for instance, one reads that "Audacity projects contain a file (MyProject.aup) plus an associated data folder (MyProject_data) full of hundreds or thousands of audio files." There is no explanation (or link to it) about why a project should have hundreds or even thousands of audio files associated to it. So, when talking to impatient users, designers are sparing in details, and use hyperlinks to provide further explanation on demand. However, linked information is not always packaged for the *impatient*, and the change in style is clear, as is the case with further explanations about the draw tool. The text says (hyperlinks underlined): "When zoomed in to maximum level, [the draw tool] lets you adjust the volume level of individual audio samples. It can be used to eliminate narrow clicks and pops in audio by smoothing out the contour of the samples, so that one sample is not at a very different vertical position to its neighbors." If the reader follows the samples hyperlink, the explanation on the linked page begins like this:

> All sounds we hear with our ears are pressure waves in air. Starting with Thomas Edison's demonstration of the first phonograph in 1877, it has been possible to capture these pressure waves onto a physical medium and then reproduce these later by regenerating the same pressure waves. Audio pressure waves, or waveforms, look something like this: <wave length graph is shown to illustrate the point>. Audacity ©1999–2008

Before moving on to the other elements of the metacommunication template, which we are instantiating at this step of our semiotic inspection, we would like to point out that the designers' messages express a fragmented view of *who the users are*. The communication structure does not help the users navigate consistently throughout discourse that *is meant* for them. As shown in the examples above, users are very likely to encounter discourse that *is not meant* for them. In our interpretation, users are expected to behave more like information explorers and miners than communication interlocutors, which in itself is probably one of the most evident assumptions about "who the users are."

"What I've learned you want or need to do, in which preferred ways, and why."

The tasks that Audacity users supposedly need or want to do are all accessible through navigation from the "Welcome Message" (see Figure 4.2). In addition to the basic ones that constitute the focus of our study, there are others like generating effects and analyzing sound. The special interest of inspecting this fragment of the metacommunication template lies in the assumptions about the users' preferences.

As mentioned before, there is a section of documentation "for the impatient" (see Figure 4.4). All communication in this section is structured around the project window, with links to explanations about the various toolbars and other graphical representations of objects appearing in Audacity's interface. There is not, however, a single link or image referring to Audacity's main menu bar, which offers eight menus: File, Edit, View, Tracks, Generate, Effects, Analyze, and Help. The message communicated by the designers to the *impatient* is that the content conveyed through

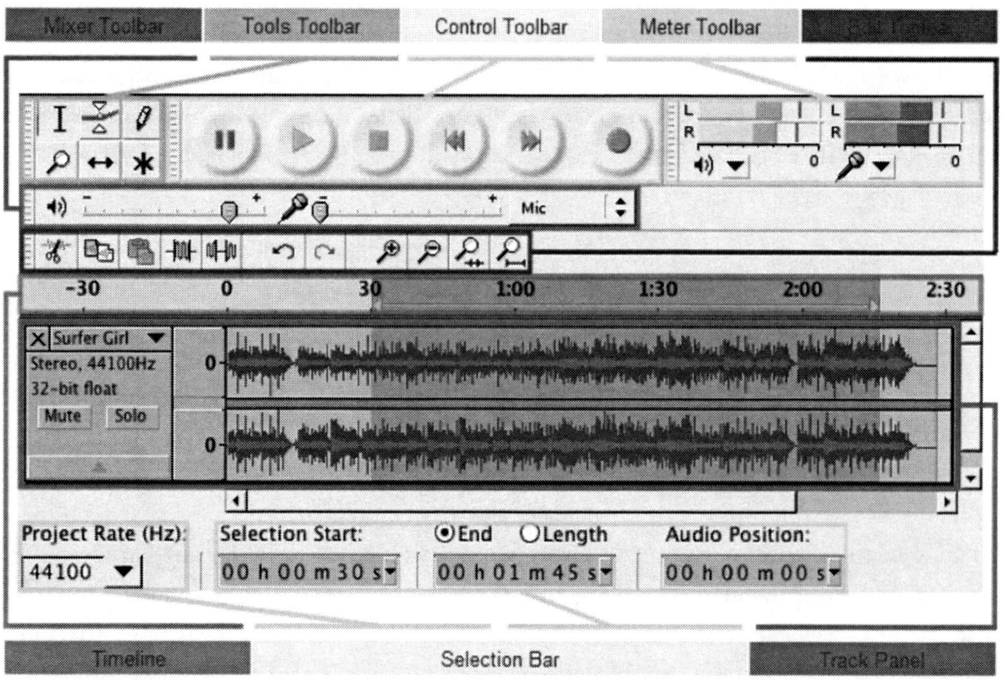

FIGURE 4.4: "Audacity for the Impatient"—Audacity's quick guide online.

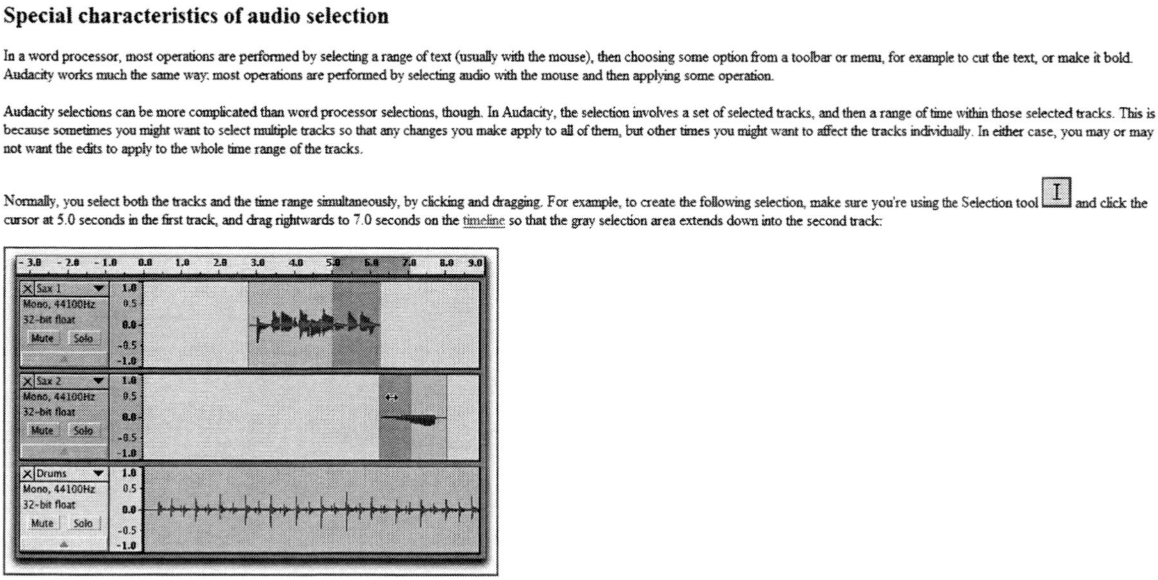

Special characteristics of audio selection

In a word processor, most operations are performed by selecting a range of text (usually with the mouse), then choosing some option from a toolbar or menu, for example to cut the text, or make it bold. Audacity works much the same way: most operations are performed by selecting audio with the mouse and then applying some operation.

Audacity selections can be more complicated than word processor selections, though. In Audacity, the selection involves a set of selected tracks, and then a range of time within those selected tracks. This is because sometimes you might want to select multiple tracks so that any changes you make apply to all of them, but other times you might want to affect the tracks individually. In either case, you may or may not want the edits to apply to the whole time range of the tracks.

Normally, you select both the tracks and the time range simultaneously, by clicking and dragging. For example, to create the following selection, make sure you're using the Selection tool [I] and click the cursor at 5.0 seconds in the first track, and drag rightwards to 7.0 seconds on the timeline so that the gray selection area extends down into the second track:

FIGURE 4.5: An excerpt of Audacity's online manual explaining how to use selection controls.

menu options is for some reason less important than, or subsidiary to, the content conveyed by visual objects on the project window. Such being the case, we find that the assumption about *impatient* users is that they prefer to interact with visual objects—namely, that they prefer direct manipulation interfaces—rather than menus. Explanations about menu content come in the "Reference" section of the online manual, or in tutorials for beginners.

Another important communication regarding the users' preferences has to do with the quality of the edited and the various controls that can or must be used to achieve it. In Figure 4.5, we see a snapshot of Audacity's online manual. In it, there are two pieces of communication that deserve our attention in a semiotic inspection. One refers to the metaphors that users are assumed to bring to bear when using Audacity. As mentioned previously, there is an explicit comparison with word processing and the patterns of interaction that users typically apply in that context.

However, users are warned that selection in Audacity "can be more complicated than word processor selections," first because there are typically many tracks in audio files—and selections may or may not involve them all—and second because selections, although spatially marked (by grayed out rectangular areas), are actually continuous timespans along the "timeline."

The "selection" entry in the online manual is possibly the longest one of all. There are eight sections in the text, each corresponding to a different form or aspect of selecting audio material for editing operations:

1. Selecting using the mouse;
2. The selection bar;
3. Selecting using the keyboard;
4. Selecting while audio is playing;
5. Snap-to guides;
6. Track selection using the mouse;
7. Menu commands for selecting; and
8. Label tracks.

Explanations about each of the above express the complexity of audio editing and the quality standards that users can expect Audacity to help them achieve. For example, the level of precision in positioning the cursor has a great impact on the quality of the result. Therefore, the designers allow the user to have very fine control of cursor positioning:

> Selection is displayed in units of hours, minutes, and seconds. As shown, it does not display any more accuracy than that, so it's hard to tell if you have half of a second selected, which can be very important sometimes. That's no problem, because Audacity gives you a plethora of choices of possible ways to display the time in the selection bar. To get these choices, click on the right triangle to the right of each box, or alternatively, right click anywhere in the box to open the context menu. Audacity ©1999–2008.

To conclude this portion of our analysis, we can say that Audacity designers communicate that users may want to do audio-editing tasks that involve fairly complex interactions with different

The Problem

Audacity has many features, but this can make it intimidating for new users. We've had many requests for a simplified version of Audacity that is easier to use. One problem is that different people have different ideas of what is "simple".

A second problem is that many settings are most useful when getting Audacity set up - settings for recording and playback, to get the MP3 library installed, settings for latency, quality and directories. These settings need to be available during setting Audacity up. They are probably best hidden from users who want to use an already set up version of Audacity.

The Solution

Audacity already contains a flexible system for changing the text in Audacity. This is used for translating Audacity into different languages. When you select a different language in the Interface Preferences Audacity will use text from the file you specify in place of the text that it has built in. The language files end with the suffix ".mo".

- We have added a small addition to this feature. If the translation of an item in a menu starts with an "!", Audacity will leave that menu item out when using that language. This allows us to radically cut down the menu.

We've created a "simplified" language file which reduces the menus in a way which we think is helpful.

FIGURE 4.6: An excerpt of the online manual section about "Simplifying Audacity."

kinds of visual representations of audio elements and operators. To facilitate interaction, they use direct manipulation in the system's interface, stressing the differences and similarities between audio editing and word editing. However, they also provide "a plethora of choices" to support the kind of precision control that is required for high-quality audio editing.

> "This is the system that I have therefore designed for you, and this is the way you can or should use it in order to fulfill a range of purposes that fall within this vision."

The communication about the system's description and use instructions occupies most of the online manual, and consists basically of 'how-to' text, which we will not repeat here. A noteworthy exception, however, is a section called "Simplifying Audacity," where designers explicitly talk about known design and use challenges (see Figure 4.6). In it, the designers tell us that they have received "many requests for a simplified version of Audacity" and that "the problem is that different people have different ideas of what is *simple*." Their solution has been to allow for extensive customization of menus, making extensions to a localization mechanism that was in place for interface language selection:

> Audacity already contains a flexible system for changing the text in Audacity. This is used for translating Audacity into different languages. When you select a different language in the Interface Preferences Audacity will use text from the file you specify in place of the text that it has built in. The language files end with the suffix ".mo."
>
> We have added a small addition to this feature. If the translation of an item in a menu starts with an "!," Audacity will leave that menu item out when using that language. This allows us to radically cut down the menu. Audacity ©1999–2008.

The idea expressed by designers is that simplification results from subtracting items, that is, hiding toolbars and menu options. This may certainly facilitate the reading of the interface, as can be seen if we compare Figure 4.7 and Figure 4.8. The latter is the result of applying step-by-step instructions provided by the design team for simplifying Audacity's interface.

However, in terms of metacommunication, there are some potentially important consequences of interacting with the simplified interface. First, we notice that signs communicating how to select audio (the selection bar " 

" and the selection tool " **I** ") have disappeared from the project window. The users' options are then to guess how to select audio via direct manipulation (see the next two steps of our inspection) and to use menu bar options (Edit, Select {All, None, Left at Playback Position, Right at Playback Position, Track Start to Cursor, Cursor to Track End}) or to use keyboard shortcuts. None of these options is particularly well communicated

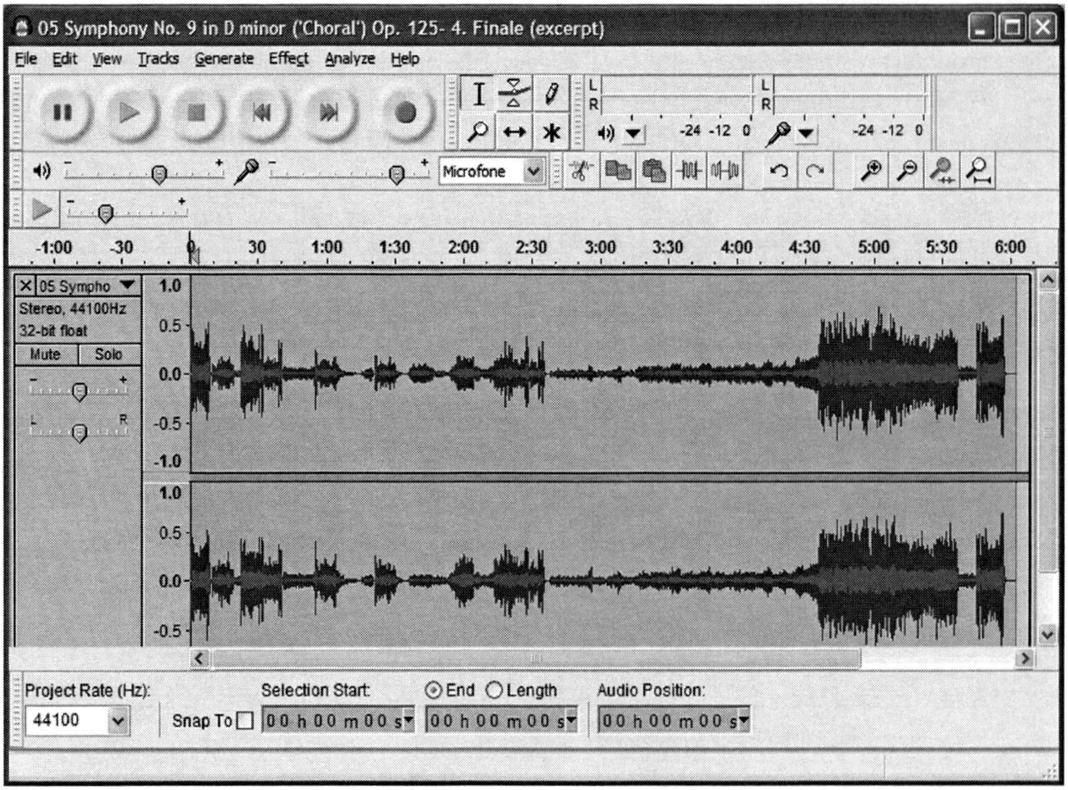

FIGURE 4.7: Audacity's default interface.

through static or dynamic signs, which means that users who do not understand the interface as it is must necessarily resort to online help. Second, the customization itself, as is often the case with customizable interfaces, is a complex process that users who already find it difficult to interact with the *default* interface are not likely to be able to carry out by themselves. In fact, Audacity's designers are well aware of this fact—they explicitly direct their communication about simplifying Audacity to "teachers and people who want to create an easier version of Audacity for others to use." And third, it is not clear how the simplified interface impacts the evolution of the users' expertise with Audacity. For example, if they do not see the selection bar, will they realize that they can make highly precise operations on audio tracks that add substantially to the quality of the editing? Or will they struggle with positioning the cursor at the right spot themselves, using direct manipulation? If they choose the latter, how will they rate Audacity in terms of the quality of the product it delivers?

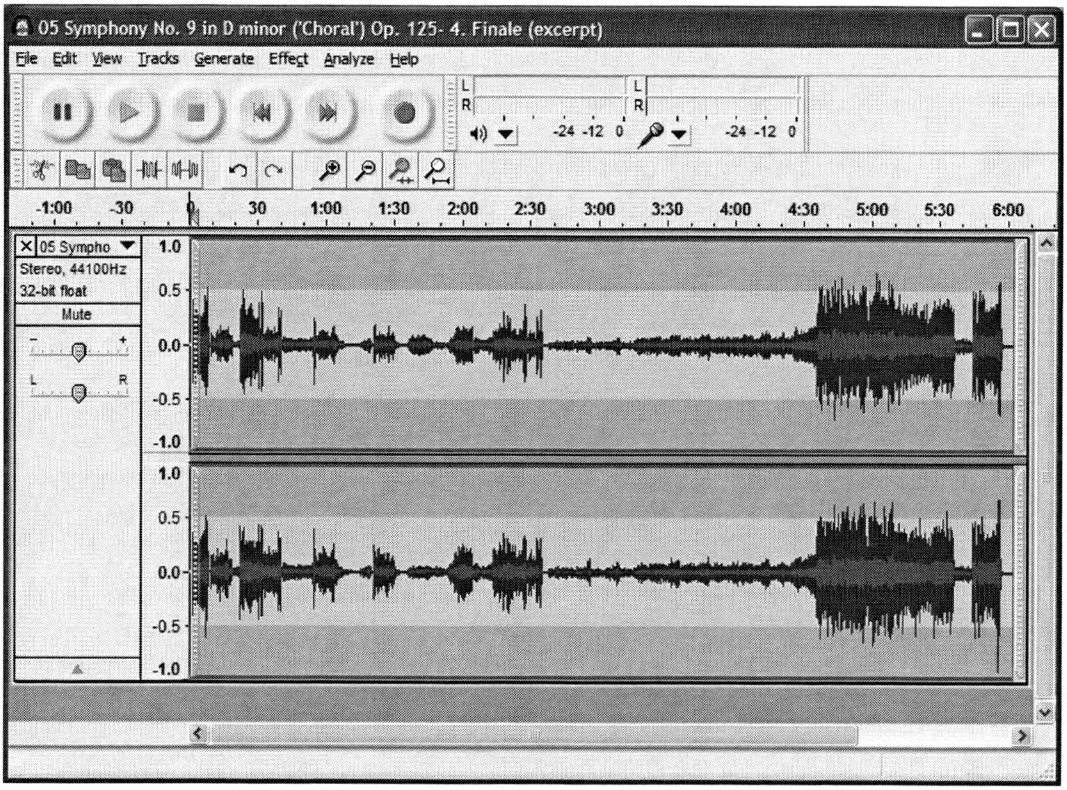

FIGURE 4.8: Audacity's simplified interface.

Before we move on to the next steps of our inspection in this study, we should remark that the considerations above refer to the *consequences* of following the suggestions about how to benefit from their design vision—not to the communication of this vision through metalinguistic signs. In this particular respect, Audacity is remarkable, giving us deep insights into the process of designing and developing a complex tool that can certainly help a very wide range of users to do audio-editing tasks that only commercial audio editors would otherwise allow them to do.

4.1.3 Analysis of Static Signs

The analysis of static signs takes as input instant representations of interface components like screen layout, menu, and toolbar structures, etc. The interpretation of these allows us to reconstruct the metacommunication message, as was done with metalinguistic signs in the previous section.

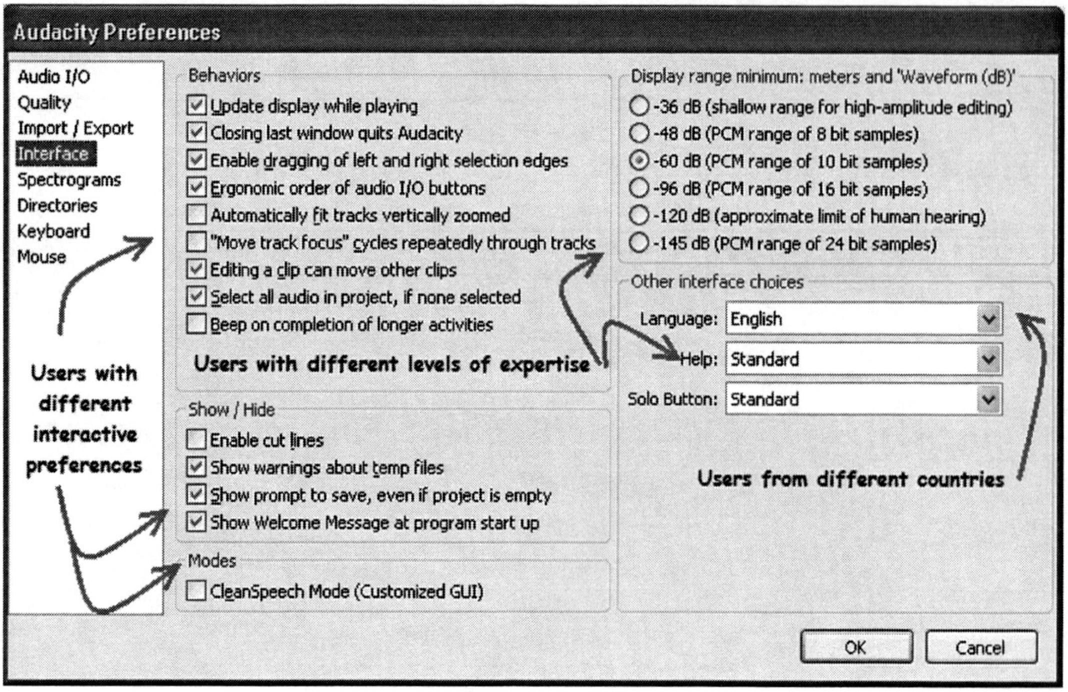

FIGURE 4.9: Interface customizations for different user profiles.

"Here is my understanding of who you are."

The reading of static signs highlights some other aspects of the designers' assumptions about Audacity users. The default screen layout (see Figure 4.7) communicates that Audacity users are assumed to be familiar with audio recorders and invited to bring this familiarity to bear when interacting with Audacity (see the playback, pause, stop, record, and other controls on the top left corner of the window).

Additionally, it is clear for all users that Audacity has been designed for a wide range of users (in agreement with the findings of our previous step). Some evident signs of different user practices and preferences can be seen in Figure 4.7, where the user is exposed to terminology that clearly signifies different levels of expertise (e.g., "32-bit float," "Project Rate (HZ)," and "Analyze" compared to "File," "Edit," "View," and "Help"). Figure 4.9 contains additional illustration of the kinds of signs that express the diversity of targeted users, even if they are to be found in the "Preferences" menu (embedded in the "Edit" menu).

Other basic assumptions about users have to do with computer literacy, especially in view of the designers' suggestions about how to simplify the interface (see Figure 4.9 and material referring to the analysis of metalinguistic signs). The simplified version of the interface clearly assumes that users are totally familiar with direct manipulation interfaces, given that the most basic operation in this style of interaction—selection—is not explicitly communicated through static signs visible in the project window.

"What I've learned you want or need to do, in which preferred ways, and why."

The communication of tasks through static signs raises interesting communicability issues. In Figure 4.10, we see the structure of all menus accessible through the menu bar. Notice that there are no menu entries for some of the most basic operations in Audacity like play, record, pause, and stop. In addition, generic terminology like "open," "save," "cut," "copy," and "paste" (expressed as verbs) is mixed with specialized terminology that nonspecialists probably cannot interpret appropriately like "labeled regions," "find zero crossings," "apply chains," and "edit chains" (expressed in various kinds of linguistic combinations).

Furthermore the categorization of action choices appearing under "File" and "Edit" is confusing. For example, under "File" there is a choice named "*Edit* Chains"; likewise, under "Edit" there is a choice named "Region *save*." Moreover, "silence" seems to be a particularly critical concept to understand, since it appears in combination with four other concepts: "edit," "generate," "analyze" (in the "silence finder" choice), and "effects" (in the "truncate silence" choice).

So far, we see that the communication about what the designers think the users may want to do shows a number of inconsistencies. However, the "how and why" they want to do it, when communicated through static signs, is consistent with the message conveyed through metalinguistic signs. The predominance of visual representations clearly communicates that designers expect users to prefer direct manipulation over other styles of interaction. However, an embedded dialog in "Preferences," just like the shortcuts associated to menu entries shown in Figure 4.10, suggests that designers have provided alternatives for users that want to interact with the system using mainly the keyboard (see Figure 4.11).

Also, the communication of precision selection mechanisms that support high-quality editing, as mentioned in the analysis of metalinguistic signs, is consistently conveyed in the default interface (see Figure 4.7). However, the relations between system states (or *contexts of interaction*) and editing tools to produce high-quality results is not well communicated. Part of the problem emerges in the analysis of static signs, but most of it becomes clear and critical only when we analyze dynamic signs.

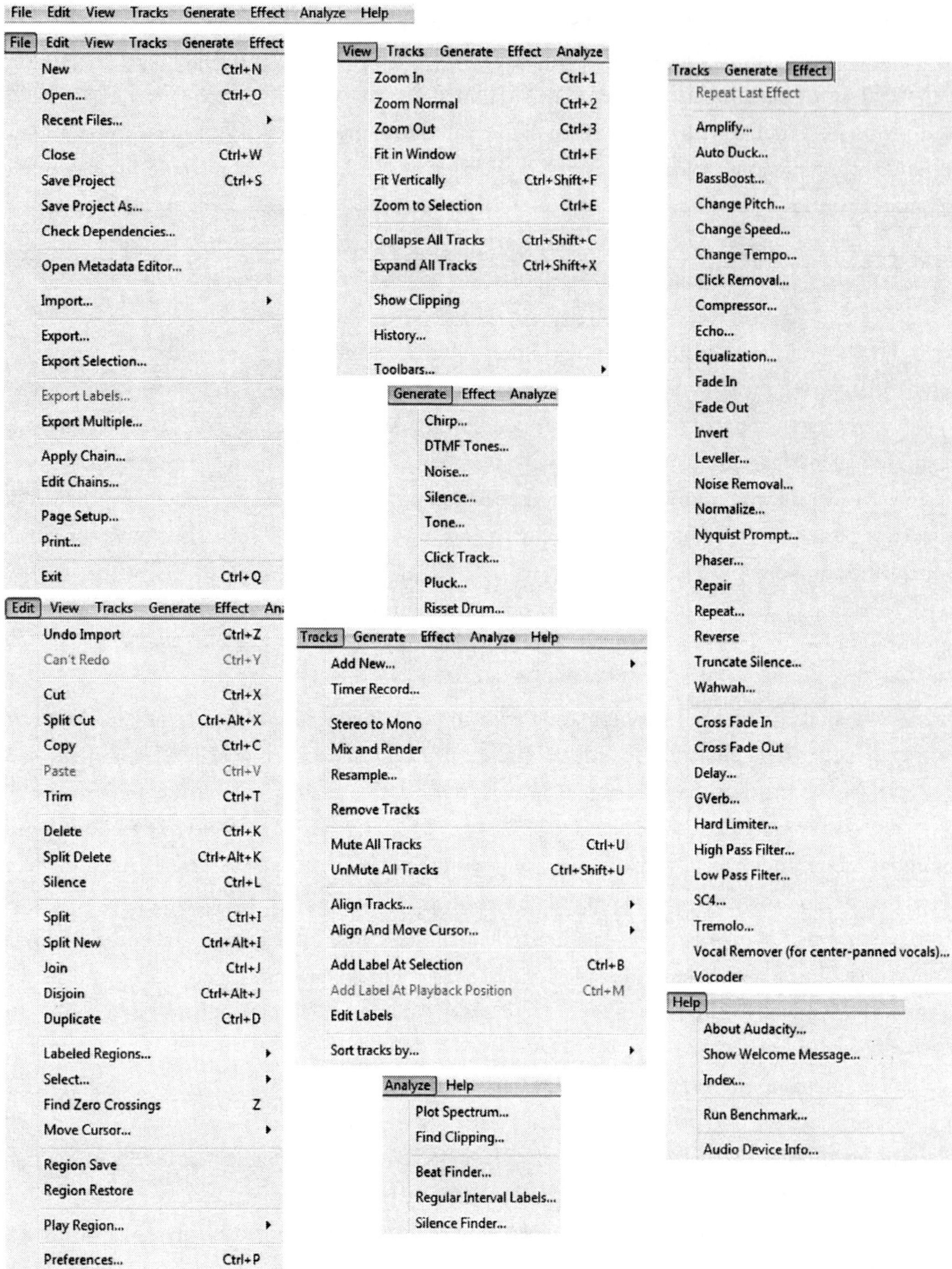

FIGURE 4.10: Audacity's menu options.

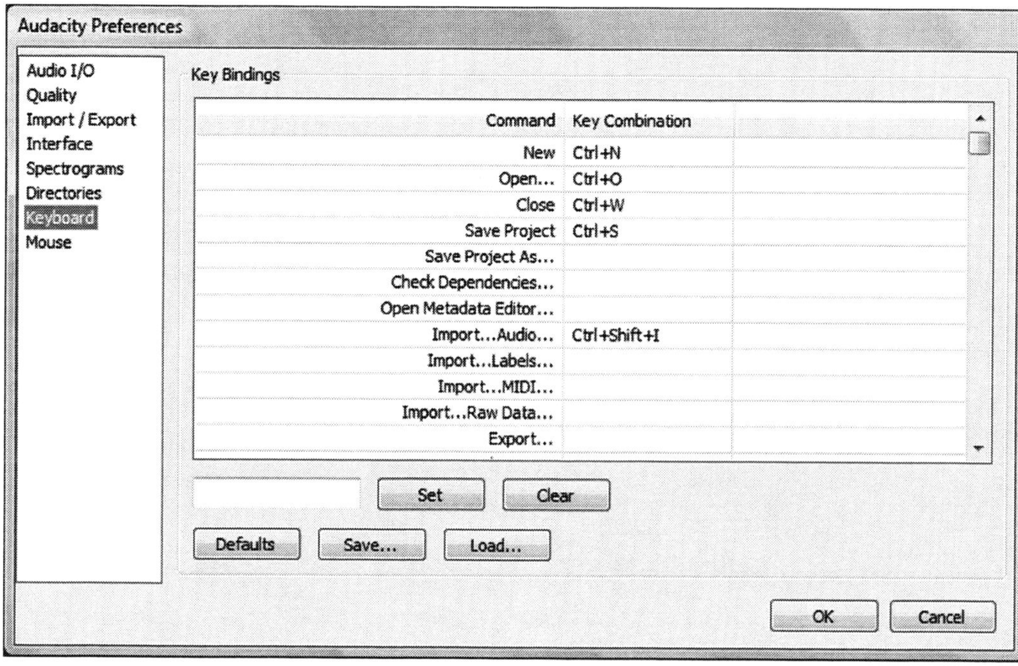

FIGURE 4.11: Audacity's alternatives to customize keyboard shortcuts.

> "This is the system that I have therefore designed for you, and this is the way you can or should use it in order to fulfill a range of purposes that fall within this vision."

The communication about how Audacity can or must be used to achieve audio-editing tasks—focusing mainly on basic tasks, given our research question—is also problematic. The analysis of menu structures above has already pointed at the kinds of ambiguities and inconsistencies we encountered in the designers' communication about the options that users must choose when trying to achieve certain kinds of tasks. Similar ambiguities and inconsistencies arise when we look at other static signs like the screen layout, for example.

In Figure 4.12, we highlight the *static* signs involved in expressing aspects of playing audio files. Element "1" indicates that the "play" button is selected. Element "2" indicates the "output volume." Element "3" indicates the playback speed (notice that the positioning of 1:1 playback speed is not clearly signified; there are no labels associated to the scale displayed on screen). Element "4" is a period of time selected on the timeline (this is the portion of audio that is going to be played; notice that the selection along the timeline, 20 to 1:15, is not coextensive with the selection on the audio

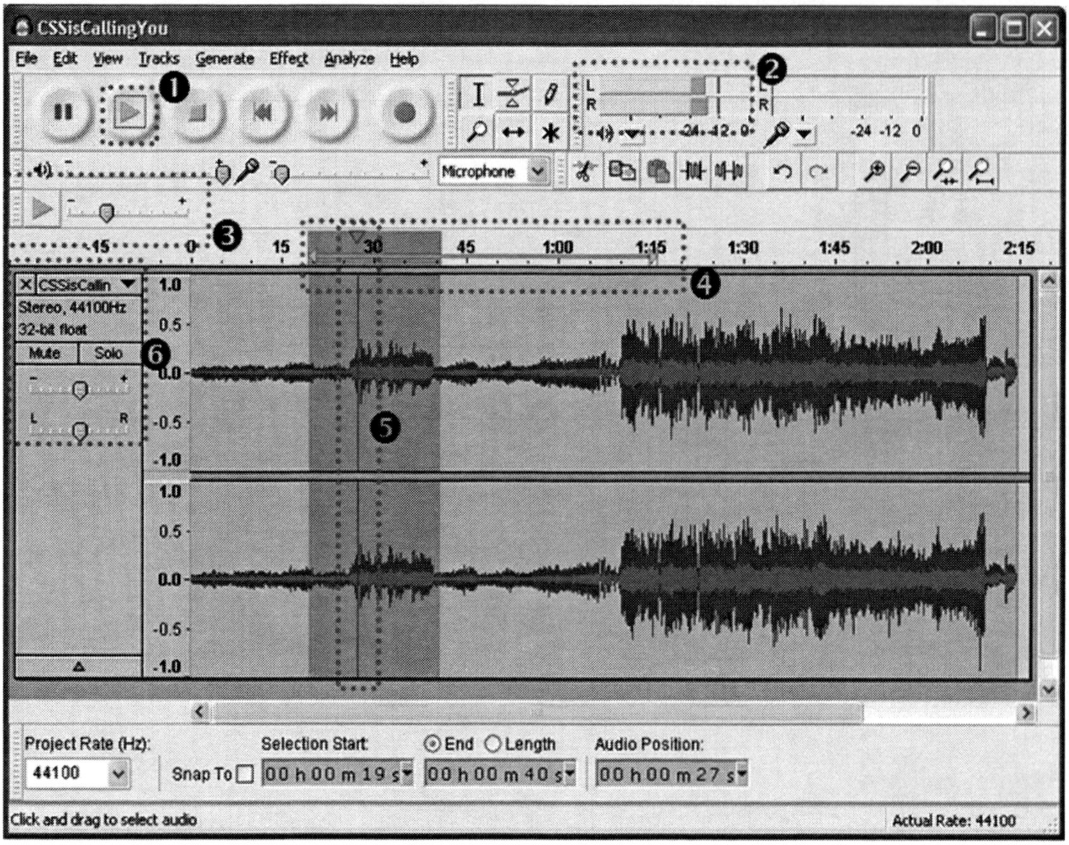

FIGURE 4.12: Visual elements involved in the signification of playback tasks.

tracks, 19 to 40). Element "5" indicates the exact playback position (the *cursor*). Lastly, element "6" indicates the "solo" tracks that are playing back at the time.

The segmented analysis of the static signs above is an important methodological step, although it should be noticed that the most salient signs expressing that the system is playing back audio are of course *dynamic*: the sound coming through the speakers and the cursor moving forward on the tracks. Nevertheless, there are at least two issues with static signs that impact the communicability of Audacity's interface. First, there are two "play" signs very close to each other, with similar appearance but considerably different meanings (see "1" and "3"). The "▶" in "1" activates the play function, whereas the one in "3" activates the "transcription mode," where the speed of playback can be made slower or faster (sliding the control toward the left end or the right end of the scale next to "▶"). In the default speed (1:1), pressing "1" or "3" launches the playback function, contributing

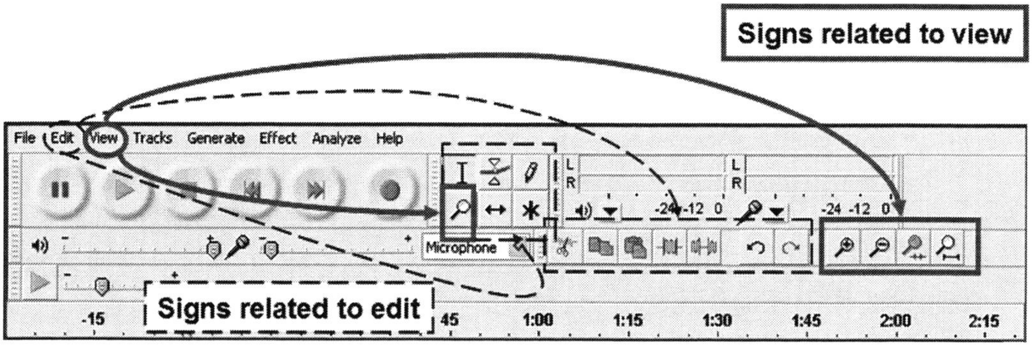

FIGURE 4.13: Visual elements involved in the signification of editing tasks.

to confuse the inadvertent user, who may then fail to learn that "3" is actually a transcription tool. And second, the visual cues regarding what is selected in this particular context of activity are quite misleading. Whereas the most salient *visual* static sign is the grayed out area on the audio tracks, suggesting that this is the portion that is going to be played, the active playback selection, and the only one to affect what the user is going to listen to, is signified by the arrow-headed grey line extending from 20 to 1:15 on the timeline (element "4").

Although designers communicate explicitly, through metalinguistic signs appearing in the online manual, that timeline selections are actually a handy way to accelerate navigation over audio track spans, this communication through static interface signs is lacking. Notice that there is no sign associating timeline selections to playback and audio track selections to editing, which is an important principle that users must understand in order to interact with Audacity in a productive way.

Although playback is a crucially important task for Audacity users, there is no doubt that editing is the heart of the system. Hence, metacommunication about what is involved in this activity, which interface elements must be activated, how and why, deserves special attention. Besides selection aspects already discussed with respect to Figure 4.12, there are also important aspects regarding the communication of combined functions that must be activated when editing audio tracks. In Figure 4.13, we show the default layout of Audacity's interface, highlighting that the spatial organization of static signs only very subtly suggests that "Edit" and "View" have important interdependencies. The "tools" toolbar (element "1" in Figure 4.12) includes a selection tool, an envelope tool (allowing for "smooth volume changes over the length of a track"), a draw tool (that we have already discussed), a zoom tool, a time shift tool (that "allows you to synchronize audio in a project by dragging individual or multiple tracks or clips left or right along the timeline" and "to drag individual tracks or clips up or down into another track"), and a multi-tool (that combines all other five tools in one). By grouping edit and view tools together and providing a draw tool that works only if the appropriate zoom level is set, the designers implicitly call the users' attention to the fact that in order to manipulate certain audio elements correctly, they must be able to *see* relevant details. But, how do they know which details are relevant?

Figure 4.13 also suggests that the static signs corresponding to elements "2" and "3" correspond, respectively, to the "Edit" and "View" menus, which are communicated as two independent functions in the menu structure. However, once again, there is a subtle sign that certain edit functions may be related to view states: elements "2" and "3" both belong to the same toolbar, which Audacity's designers have revealingly named "the edit toolbar."

The main result at this stage of analysis is that metacommunication about certain critical dependencies, such as between timeline selections and playback or between audio track selections, zoom level, and enabled editing operations, should be more clearly communicated. The implicit cues mentioned in preceding paragraphs are not likely to be captured by users that do not like to read manuals, except perhaps at the expense of extensive trial and error and experimentation with the interface.

4.1.4 Analysis of Dynamic Signs

The analysis of dynamic signs is the last segment of inspection in SIM. Following the same structure adopted for presenting our results in previous steps, we will only describe the main points in association with each portion of the metacommunication template.

"Here is my understanding of who you are."

Dynamic signs reinforce the communication that users are familiar and comfortable with direct manipulation interfaces. Most of the action in Audacity is visually codified and commanded with the mouse. Although there are keyboard alternatives for commands, changes in cursor shapes and salience of visual representations of both objects to be edited and tools that can be used to edit them clearly communicate the importance given by designers to Audacity's graphical interface.

It should be noted at this point that one of the prominent features of a direct manipulation style of interaction is that, unlike menu-based or command–line interaction, there are no verbal cues to help users guess the spectrum of opportunities associated to certain interface actions. For example, the existence of two menu options like "Copy" and "Duplicate" in the "Edit" menu (see Figure 4.10) immediately communicates that users might want to know about "duplications" (which in English are semantically defined as "copies") precisely because their copresence in this menu tells that they are not synonymous. However, this association is not communicated through direct manipulations of audio tracks in Audacity. Hence, a user that limits herself to using only the visual representations appearing in the project window will probably miss some of Audacity's nice features for editing audio. In this respect, it is useful to recall that Audacity's quick guide "for the impatient" stimulates only interaction through direct manipulation.

> "What I've learned you want or need to do, in which preferred ways, and why."

Regarding this portion of the metacommunication template, dynamic signs communicate important aspects of Audacity's design. One of them is the effect of customizations. For example, as shown in Figure 4.9, users can choose to configure the interface with the "ergonomic order of audio I/O buttons." Users are not likely to know what this means, and the most direct way to find out is to experiment: to choose this configuration and see what it looks like.

Another aspect highlighted by dynamic signs is how the designers use image and sound to provide feedback and try to give users an exact notion of what they are doing with audio. Hence, in connection with direct manipulation preferences, there is an implicit communication about how users are expected to relate auditory and visual signs, sound patterns with waveforms and volume meters. The ability to map visual manipulations onto operations with audio objects is crucial for productive interaction with Audacity, and animation is extensively used to help users to this mapping correctly.

> "This is the system that I have therefore designed for you, and this is the way you can or should use it in order to fulfill a range of purposes that fall within this vision."

At this final step of segmented analysis in SIM, the dynamic signs in Audacity's interface reveal some important communicability issues. Two of them have to do with the communication of why certain asymmetries are in place; others are related to the communication of enabling contexts for action.

The first asymmetry we want to discuss is that involving open/save actions, on the one hand, and import/export, on the other. The primary meaning of "open" is to "open an Audacity project." Through metalinguistic signs, Audacity designers extensively explain that Audacity projects have their own format and that if users want to work with (or generate) audio files in formats like "wav" or "mp3," for instance, they must "import" (or "export") the project file, commanding explicit conversion operations. However, dynamic signs undermine this communication in an important way. For example, Audacity "opens" mp3 files—the system automatically converts the file into Audacity's internal format. Thus, there is virtually no difference between the command "open" and the command "import," which may suggest to users that the same kind of automatic conversion will be performed by the system in the dual circumstance. For instance, if the user "opens" (instead of importing) a "wav" file and, after editing it, decides to "save" it, it is perfectly sensible to expect that the system will "export" it to (or "save it as") a "wav" file (i.e., automatically convert it back to "wav"). However, this does not happen in Audacity. "Saving" and "exporting" remain separate things, whereas "opening" and "importing" have been merged.

Another important asymmetry is found when manipulating waveforms on tracks and timespans on the timeline. When users select a portion of an audio track, this selection automatically entails an explicitly marked selection on the timeline. This communicates that there is solidarity between spatial manipulations (i.e., selections of regions on tracks) and temporal manipulations (i.e., selections of period of time corresponding to the duration of the selected region), which is true only if the user performs the actions in this particular order. If the user selects a period of time on the timeline, there is no selection of the corresponding region(s) on the audio tracks. As mentioned before, there is a reason for dissociating temporal and spatial manipulations in this case: that is, temporal selection actually determines what the user will listen to through playback, but not necessarily what the user wants to select for further editing. The user may be simply searching for sound patterns down the track that are similar to the one he has currently selected and is ready to copy or apply an effect to (see Figure 4.12 for an illustration of dissociated temporal and spatial selections). This communication, however, is very precarious and confusing when expressed through dynamic signs. Contrary to the intended message, users may end up thinking that the asymmetry is a *bug* in the system, and not a (useful) feature especially designed for certain use situations.

The last kind of asymmetry is more of a perceptual nature than a conceptual one. It involves the use of sound and visual representations to convey system states that enable or disable various kinds of editing functions. For example, such ordinary things as cut, copy, and paste audio can only be done if the playback is *stopped.* The most salient communication that this is the case is that the user stops listening to what was playing—for all practical purposes, no sound coming from the speakers, combined with a still image on screen, is a clear sign that the playback is *stopped.* However, there are two system states associated to this strong perceptual state experienced by the user: one is the result of pressing the "stop" button (in which case the playback *stops*), and the other is the result of pressing the "pause" button (in which case the playback *does not stop*). The communication of feedback in this case is problematic because the most important cue is *visual,* although the most salient sign is *aural.* When the playback is paused, the visual representation of the ongoing playback is "suspended"—the cursor position is retained, the playback span on the timeline is marked, etc. However, when the playback is stopped, the visual representation changes more drastically—the playback cursor and timeline selection disappear, and the visual representation retains only spatial selections seen during the playback (if any). Other visual cues are the states of playback control buttons themselves. When the system is paused, the "pause" button looks *pressed.* However, when the system is stopped, the "stop" button is simply disabled. This additional asymmetry introduces ambiguities that get in the way of an already complex piece of metacommunication that designers are trying to achieve.

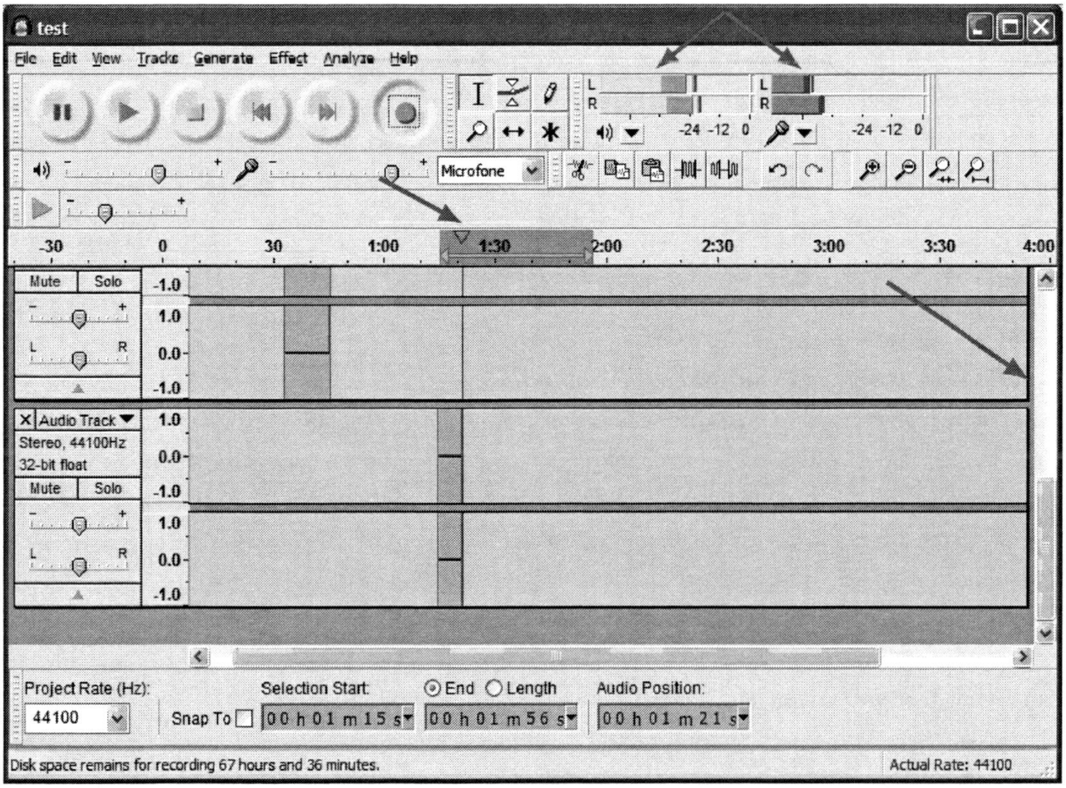

FIGURE 4.14: Visualizations of aural signs in Audacity.

A last observation about dynamic signs, also related to auditory and visual feedback, has to do with the signification of simultaneously active volume meters if the user has an active track in the project and decides to record something without *muting* the active track. Figure 4.14 shows a representation of one such situation. The system interface state shown in this figure is even more interesting to analyze because it does not show all the active audio tracks. As the scroll bar on the right side of the screen suggests, there are other tracks "up," although they are not visible in this particular state. However, knowing that they are there, and realizing that they are actually *being played* as the user records something on the microphone, is essential for the correct interpretation of both active volume meters on top of the screen. The one to the left corresponds to the volume level of playback (output), whereas the one to the right corresponds to the volume level of recording (input). Dynamic signs direct the user's attention to the progression of recording on the last track (at the bottom), which may be misleading for novices (and occasionally for experienced Audacity users, as well). It is fairly difficult to interpret, without hesitation, if the recorded material includes the playback of other tracks or not, and what the output volume meter is actually *telling* in this particular configuration of the system. The pragmatic interpretation of dynamic signs may become even more difficult if we think of actual use situations where a user is holding a microphone in front of his mouth and saying something that he wishes to record (see our inspection scenario for an illustration). His attention is likely to be directed primarily to the recording activity, which presents the risk that the user will not listen to the other tracks playing back, and thus not realize that his recording is coming out wrong if the idea is to move the recorded piece to some other place along the timeline *without* a trace of other tracks' contents at recording time.

4.1.5 Collating and Comparing Analyses

The purpose of comparing metacommunication as expressed by metalinguistic, static, and dynamic signs in isolation is to identify what the designers are telling the users in each case and to analyze how they use and combine the three classes of signs to compose their global message to users. In this case study, it is important to keep in mind that our focus of investigation are the *strategies to communicate a system's basic functions and how they can or should be used.* This study with Audacity should thus help us to find an answer to a research question that clearly transcends the specificities of this system's interface. Consequently, at these last two steps of SIM we will try to show how we can articulate more general knowledge from specific evidence collected so far.

We begin with specific findings that the "collate and compare" stage of our semiotic inspection of Audacity has produced. In all three preceding steps of analysis, we found communication saying that the designers assume and expect that users are familiar and comfortable with direct manipulation interfaces. We also found, when analyzing metalinguistic and static signs, that Audacity has been designed for a wide variety of users and that there are certain mechanisms in place to attend to different user profiles and use situations envisioned by developers. In the analysis of dynamic signs, this sort of communication was less salient than communication about how the system works, the effects it achieves, the various contexts or interaction, and the like.

Moreover, the analysis of metalinguistic signs gave us important insights about the context of design and development of Audacity. We learned that this is a system developed by a team of volunteers and that there is active participation of users in reporting their experience and requesting new features. In particular, we found communication telling us that Audacity developers have been asked to *simplify* the system, and that this is a challenge for them because "different people have different ideas of what is *simple.*" In an attempt to resolve the problem, at least partially, the designers offer advanced users the opportunity to customize Audacity's interface and hide certain elements that may add unnecessary complexity to interaction. They also give step-by-step instructions, which can be followed even by first-time users, to take many elements away from the interface and have a *cleaner* interface to interact with.

Although the conditions of software production are not explicitly signified through static and dynamic signs (except in combination with metalinguistic signs, as is the case of contents in the "About Audacity" window, accessible through the "Help" menu), there are some important traces of it in the other two contexts of analysis. In the analysis of static signs, categorization problems spotted in the "File" and "Edit" menus, for instance, signify that the design and development team probably does not have (or does not use) a support tool to register general design principles that should guide decisions about where and how to include new features in Audacity's interface. It is also revealing to find the option to customize the interface using the "ergonomic order of audio I/O buttons." Calling the alternative order *ergonomic* and not having it as the default choice (but as the result of deliberate customization) is a clear sign of competing design views in the development team. In general, *ergonomic* designs are the natural choice for all interfaces, and there is no apparent reason why a *non-ergonomic* design should be favored in Audacity's default configuration. So, this curious customization choice actually expresses that the meaning of *ergonomic* is not a consensus

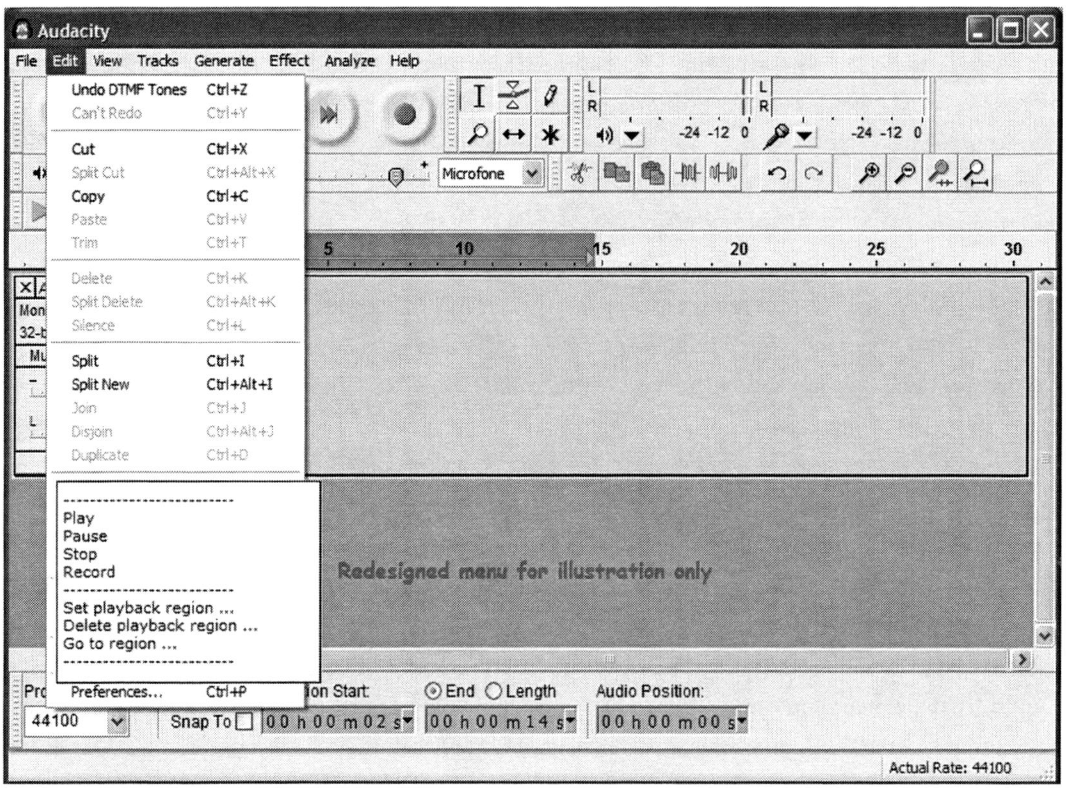

FIGURE 4.15: Redesigning metacommunication in Audacity through static signs.

among developers and directs our attention to the challenge of reaching consensus (and enforcing the consequences of it) among a team of volunteers.

In the analysis of dynamic signs, there are also traces of the challenges faced by volunteer developers. For example, the behavior of similar actions and/or their corresponding effects is not communicated homogeneously to users. Narrative cursors used to communicate the possibility to change selected areas on the timeline and on audio tracks have different shapes for no apparent reason. Likewise, similar dialogs under the "Generate" and the "Effect" menus enable different interactive opportunities. For instance, when the user considers "amplifying" selected audio, he or she can "preview" what it will sound like. However, when considering the generation of a "chirp" or "DTMF tones," there is no preview to help the user decide if this is what he or she wants.

Another important finding when comparing the results of segmented analyses is that very relevant communication about certain features of Audacity conveyed through metalinguistic signs in online help are not reiterated and reinforced well enough (if at all) by communication through static and dynamic signs. A noteworthy example is the communication about how the timeline can be manipulated. In online help material, we find out that selecting portions of the timeline is an efficient way to navigate through audio tracks, listening to different parts of it while retaining selected portions of audio tracks for editing (if so desired), or setting fixed regions to facilitate repeated playback during more elaborate editing activities. However, the value of this mechanism is not communicated through static and/or dynamic signs. In particular, it calls our attention that the words "timeline," "navigation," or "go to"—which evoke important concepts associated to the mechanism under discussion—do not appear in Audacity's most immediately accessible interface elements (i.e., the visual elements on the project window, or the top-level menus and their corresponding options). In Figure 4.15, (see offset rectangle in the "edit" menu), we illustrate the effect of signifying such concepts, sketching an example of how communication through static signs can reiterate and reinforce important communication conveyed through metalinguistic signs.

From the observations above, we already have important elements to conclude what strategies the designers have chosen to communicate the basic functions of Audacity and how they should be used. First, they use metalinguistic signs almost exclusively to communicate information that is crucially important for users to understand how Audacity works and to benefit from numerous powerful features, encountered mainly in commercial digital audio editors. Second, by favoring the direct manipulation style of interaction so strongly over menu-based interaction, they miss the chance to use static signs to induce and stimulate the users' interpretations of which manipulations they can or should try to do. Third, because digital audio editing is per se a fairly complex task even at the most basic levels of operation, involving relatively sophisticated mappings between spatial and temporal representations, as well as between visual and auditory signs, the absence of more elaborate verbal communication during interaction imposes a heavy burden on visual communication. Hence, interface design challenges are yet more difficult than they could be if menu-based interactions were more extensively explored by the design team. And fourth, the designers' intent to meet the needs of a wide range of users with a single system has, as they acknowledge in the online help, come at the expense of excessive complexity. The analysis of static and dynamic signs indicates the difficulties faced by users trying to make sense of metacommunication conveyed through Audacity's interface. Even the customization mechanism that designers have devised to alleviate the problem (an extended use of the language configuration file) is complex to use and beyond reach for a nonexpert user in need of simplification. It is interesting to notice, in this respect, that their solution restates a cultural value for communities of open-source software developers—they invite expert users to customize and distribute simplified interfaces for nonexperts, that is, those who know more may volunteer to work for those who know less.

4.1.6 The Quality of Metacommunication in Audacity

The first conclusion of this semiotic inspection is that communicability in Audacity can be considerably enhanced to overcome the challenges listed in previous sections of this chapter. Some possible alternatives have even been briefly described (see Figure 4.15 for an example). However, our main interest with SIM is oriented toward scientific research, not technical improvement. So, the question we should answer in conclusion is: *what HCI knowledge have we gained about strategies to communicate a system's basic functions and how they can or should be used?*

As is the case with qualitative methods, SIM results must not be taken as generalizations or predictions about what *will happen*, but rather as knowledge derived from interpretations of what *has happened.* So, we structure our conclusions around three topics that are central to our HCI theory:

1. Strategies for integrating metalinguistic, static, and dynamic signs to achieve intended metacommunication;
2. Alternative metacommunication templates for different user profiles, and mechanisms for switching consistently across them; and
3. Communication support at design and development time, so that cohesive and consistent interactive discourse is generated at interaction time.

Regarding (1), our conclusion is that important communication about design intent or design rationale (i.e., important passages of the metacommunication template) should be conveyed and reinforced by all three classes of signs: metalinguistic, static and dynamic. It has been known for a long time now that when having to decide between learning and guessing, users typically do the latter. In *The Paradox of the Active User* Carroll and Rosson (1987) describe the cognitive tensions when users have to choose between more *throughput* and more *knowledge*. According to the authors, when what they are doing *looks like* it is advancing their goals, users "hastily assemble ad hoc theories" of how the system works. Although their work is founded on cognitive theories, which they use to discuss alternatives for attacking, mitigating, or designing for the paradox of needing to learn more but deciding not to do so, Carroll and Rosson provide a bridge to semiotic theories such as the ones that inspired semiotic engineering.

What they characterize as a hasty assembly of ad hoc theories is what semioticians refer to as abductive reasoning. Abductive reasoning is the basis of *semiosis*, the fundamental sense-making process by which signs come into existence. In the context of metacommunication, one of the contributions of semiotic theory to semiotic engineering is to show that the users' exposure to metalinguistic, static, and dynamic signs mutually influence their interpretation of messages they get from each of them individually. In other words, the associative chains triggered by related metalinguistic, static, and dynamic signs should *signify* consistent and cohesive meanings. For example, as discussed in previous sections, if Audacity designers wish that users will understand and benefit from certain features that they have included in the system, this must be signified in online help, of course, but also be *suggested, evoked, prompted, teased*, by static and dynamic signs that users are likely to *associate* to meanings that gravitate close to the idea that designers want to communicate.

Regarding (2), we conclude that metacommunication cannot be efficiently and effectively conveyed through a single interface to users with widely different backgrounds, motives, and attitudes toward the system. The structure of communication requires that the roles of senders and receivers can be consistently assigned to *identifiable* subjects, individually or collectively. *Identifiable* is a key concept in this context, even if receivers are collectively addressed, as is the case with *mass media* communication, for instance. Notice that we naturally *identify* (i.e., we assign an identity to) collective interlocutors by referring to them as "the New York Times," or "the government," or "Audacity." By the same token, we are sensitive to discourse that violates this identification. For example, when "the government" passes one piece of legislation that protects the environment and another that creates incentives for technology that pollutes the air, we demand that authorities be consistent. Even more fundamentally, we *identify* ourselves as receivers of all communication addressed to us. Thus, we are even more sensitive to inconsistencies that we detect in the way our interlocutors are treating *us*. For example, we are easily disoriented by interlocutors that, while talking to us, treat us as partners *and* competitors, as knowledgeable *and* ignorant, or as their superior *and* their subordinate.

While interacting with systems, users behave in the same say. They view "the system" as an *identifiable interlocutor*, and expect to be recognized as one, too. They are thus disoriented when they get messages that reveal inconsistent assumptions about the communicative setting in which they are both involved. This is why communicating different things to different users *at the same time and*

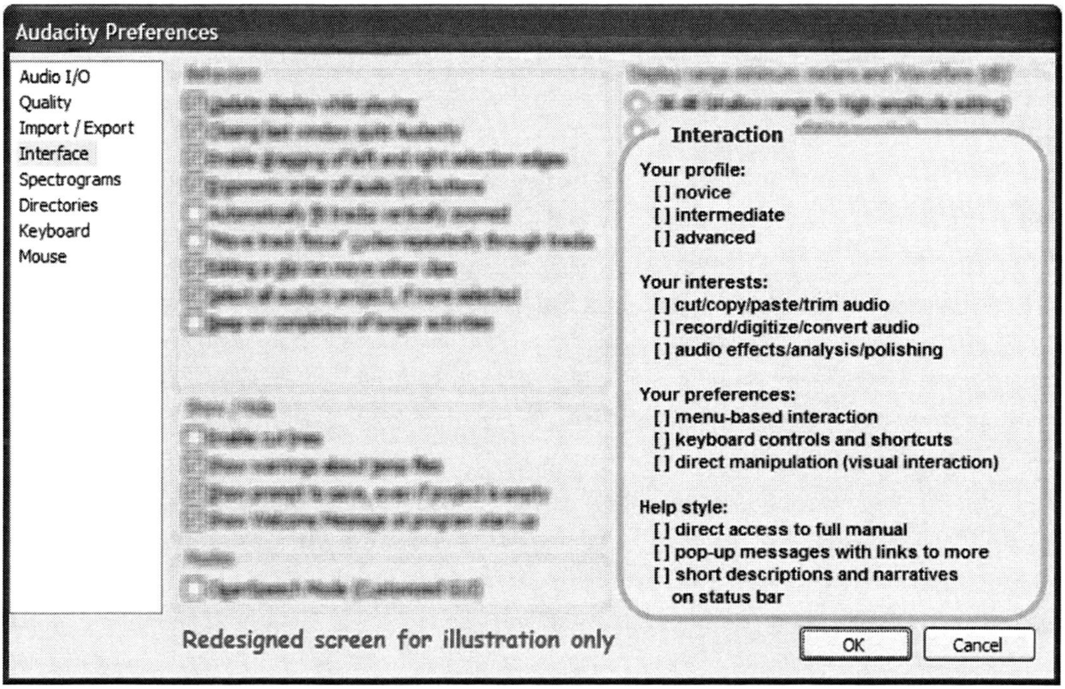

FIGURE 4.16: Communicating customizations of metacommunication strategies.

place (i.e., through the same interface) is a problematic choice in HCI. Other than in a small set of highly conventional or controlled contexts, there is probably no single piece of metacommunication discourse that will be universally adequate for a wide range of user profiles. Hence the importance of and increased demand for customizable interfaces. The problem, however, as we have seen in our inspection of Audacity, is that most such interfaces support mainly two kinds of adaptation: alternative "skins" that affect mainly visual and perceptual aspects of the design, like the *ergonomic* interface option; and activation/deactivation of modules and functions. This "compositional" approach to customization, which achieves adaptation by adding and subtracting elements of the interface, is useful in many situations but fails when what is required is not to say more or less of the same, but to say it differently, and possibly for different purposes.

So, the HCI problem that the study with Audacity helps us to unveil is the need for methods, models, and artifacts that will help designers conceptualize, represent, analyze, test, and implement different *conversations* about the system with different kinds of users—specialized metacommunication about and by means of the same piece of software. Much of the fundamental knowledge needed to achieve it has been investigated by AI researchers interested in building intelligent systems capable of handling natural language conversations (Kobsa & Wahlster, 1990; Maybury, 1993; Maybury, 2004; Moore, 1995). Part of their job is to propose models and strategies to represent the users' knowledge states, beliefs, intent and motivation, and to build inference engines that can be used to decide how the system should interpret the users' utterances and what it should say next. Although we do not advocate that this pattern of interaction is the best, or even that it is useful and usable for systems like Audacity, for example, the kinds of models and strategies used in AI can certainly help us organize meaningful conversations with different user profiles, and pre-compile them for activation at interaction time. This would allow us to think of more elaborate customizations, where instead of just selecting elements *to show and hide* the users would be able to select topics *to talk about*. As shown in Figure 4.16, users should be able to choose different metacommunication strategies depending on their level of expertise, goals, preferred interaction style, and disposition to learn more.

Finally, regarding (3), we conclude that the communicability of complex interactive messages, produced by a team of designers and developers, must be directly dependent on the ability of all team members to understand and articulate—at least verbally, if not also through system specifications and actual program coding—the instantiated content of the metacommunication template (see p. 16). In this respect, the study of Audacity, developed by a team of volunteers following free and open-source systems development practices, is particularly illuminating.

The main page in Audacity's Web site has a link named "get involved," where users are invited to give feedback and report bugs, and developers (identified as C++ programmers) are invited to join the development team and help improve Audacity. By following the corresponding links, we find resources at Ohloh ("the open-source network") and Audacity's Wiki. Although there is a "Developer Guide" online, we see that there are only very general guidelines in it, and they are mainly addressed to *programmers*, not *designers*. For example, there is information about Audacity's architecture, about how to fix bugs, about libraries and plug-ins, but no information about users, supported tasks, contexts of activity, and so on. We learn that decisions about new features and dialogs are discussed in the developers' mailing list, although it does not mention user models, task models, interface design guidelines, and the like.

An adequate *semiotic engineering* of user interfaces requires an appropriate communication infrastructure that the design team can use to discuss and decide what they are about to do and why, bearing in mind that it is *the users* that they are ultimately trying to help and attend to. Therefore, the availability of representational artifacts that can capture and express the design vision that the whole team should be trying to communicate is essential for successful semiotic engineering. Without it, the development team is lacking in *common ground*, and hence cannot possibly produce cohesive metacommunication.

Free/open-source software (FOSS) development teams make intensive use of computer-mediated communication and group technologies, and are often in direct contact with communities of users that participate in forums, mailing lists, and wikis. Barcellini, Détienne, and Burkhardt (2007) remark that such practices have produced a new kind of participatory, "pushed-by-users" design. In it, users that participate simultaneously in discussions about development issues and user feedback issues act as mediators in emerging software development practices that deserve more attention from researchers. In spite of such participation, however, usability issues in FOSS are not likely to be substantially improved at the same pace as new FOSS applications are produced (Nichols & Twidale, 2003; Twidale & Nichols, 2005). Among the challenges to be faced is the predominance of *bug reporting*. Most discussions involving developers and users center on *bugs*, which promotes a *programming* perspective rather than a *design* perspective upon user interface and interaction issues. Moreover, as Twidale and Nichols (2005) remark, certain user requests and suggestions—if accepted—have a ripple effect. They affect parts of the program that are stable and finished, whose developers may not even be in the team any longer. So, the cost of acceptance is often too high. This situation may not only create tension among developers when usability issues are raised, but also stimulate an *interface*-centered view of HCI, which misses many of the fundamental *interaction* design decisions that affect users so profoundly. The evidence we get from a semiotic inspection of Audacity goes exactly in this direction. It shows that metacommunication achieved through the interface is almost a collage of interactive discourse fragments that may be locally cohesive within the narrow scope of a specific task, but tend to be globally uncohesive in the broader context of activities related to editing digital audio.

4.2 A COMMUNICABILITY EVALUATION OF AUDACITY

The communicability evaluation of Audacity was carried out to validate the conclusions of our semiotic inspection. As mentioned in Chapter 3, in qualitative research the purpose of triangulation procedures like this one is not to *replicate* results achieved with different methods and treatments, but to verify that all results are *consistent* with each other, and that they point to higher-order knowledge elements that explain (or might explain, as a result of further research) the findings at hand. Compared to SIM, CEM has the advantage to involve user observations and to give us objective evidence of how metacommunication is received during interaction.

We should emphasize that the order of application of SIM and CEM is important in this study. A semiotic inspection carried *before* communicability evaluation helps to preserve the inspector's focus on the quality of the *emission* of metacommunication, independently of *reception* evidence collected from users. If the order is inversed, the inspector's judgment is certainly influenced by such evidence, making it methodologically more difficult to keep the right focus during the

TABLE 4.1: CEM test participants.

GROUP	PARTICIPANT'S ID CODE	PARTICIPANT'S PROFILE
Novices in audio editing	N1	Professional systems developer, musician, Ph.D. student (computer science)
	N2	Economist, casual computer user, interested in downloading music and burning CDs
	N3	Undergraduate student (History), works in a digital documentation project, experience with image processing
Participants with previous experience with audio editing	E1	Ph.D. student (computer science), professional multimedia application developer
	E2	Graduate in History, professional DJ, produces digital audio files regularly
	E3	Computer scientist (Ph.D.), musician, and composer
Audacity user	AU	M.Sc. student (Computer Science), open-source software developer

inspection. There are, however, opportunities to apply SIM *after* CEM. For example, when CEM produces puzzling results, SIM can be advantageously applied taking such results as a new focus of inspection.

4.2.1 The Evaluation Setting

We invited six participants for our tests, all first-time users of Audacity. The purpose of working with first-time users was that the *reading* of interface signs in first encounters with systems is very intense, giving us a prime opportunity to observe how metacommunication is received. We also invited a seventh participant, an experienced Audacity user, to give us contrastive evidence of metacommunication reception. The complete set of participants is presented in Table 4.1.

First-time users that participated in our experiment were divided in two groups: three novices (N*) in audio-editing activities and three people with considerable experience (E*) with *other* audio-editing systems. As shown in Table 4.1, participants had different profiles in terms of professional experience, training, and potential interest in audio files. All such profiles are consistent with the wide range of users that Audacity's developers intend to reach.

The test scenario was basically the same we used for semiotic inspection (p. 52), except that we added a sentence saying that the main character (the role played by participants) decided to save his ringtone as an MP3 file on the desktop. SIM and CEM scenarios do not have to be the same for triangulation purposes, but the possibility of doing it in our study increased the cohesion between both evaluations. Test procedures followed CEM steps presented in Chapter 3, and participants had a maximum of 30 minutes to run the scenario. This time limitation was part of the narrative defining the role that participants should play in the test.

Regarding tasks to be achieved, participants had no difficulty with opening Audacity, loading an MP3 file, or listening to the audio. However, the only participants that were totally successful in running the entire scenario, were E1, E3 and AU (see Table 4.1). N1 and N3 successfully recorded their message, adding it to the ringtone, but failed to generate the MP3 file in the end. N2 had so much difficulty trying to find the appropriate functions to manipulate the audio file (select portions of it and clip it as desired), and to record his message, that he gave up trying to run the scenario altogether. E2 had no difficulty to select and clip the audio, and did it fast (less than 5 minutes). However he simply could not find out how to record his message. He spent more than 20 minutes trying to do it, and eventually gave up.

4.2.2 Tagging and Interpretation

In principle, the time to accomplish tasks is not important for us. The activity involves aesthetic aspects like choosing portions of an audio file that will *sound good* in a ringtone, clipping audio with enough precision so that different parts *blend in* nicely, and so on. Some participants were clearly enjoying the activity and taking their time to produce something they really *liked*. N1, for example, asked us to send him his ringtone by email because he wanted to show it to somebody else. E3 told us that he "could have spent days" producing a ringtone that would sound "just right."

Time to accomplish tasks *is* significant, however, when associated with communicative breakdowns. For instance, E2 spending 21 minutes and 20 seconds trying to figure out how to record a message using a microphone properly connected to the computer is a clear indication that something is wrong with the designers' communication with this user. Likewise, the time it took N2 to give up trying to run the scenario (23 minutes and 53 seconds) is indicative that this participant made a considerable effort to get the designers' message and accomplish the proposed tasks. In both cases, the fact that N2 and E2 were doing a lab test (i.e., an induced activity in an artificially

TABLE 4.2: Strategies for integrating metalinguistic

	N1	N2	N3	E1	E2	E3	AU
"Where is it?"		*			*		
"What happened?"					*		
"Why doesn't it?"	*	*	*	*	*		
"Looks fine to me."	*		*		*		
"Where am I?"	*	*	*	*	*		
"What now?"		*	*	*			
"Oops!"		*	*	*		*	
"I give up."		*			*		
"I can do otherwise."				*		*	*
"I can't do it this way."	*						

created context of use, however plausible) must be taken into account. E2 explicitly said in the post-test interview that he had spent that much time trying to find recording function "because it was a test." In real life situations, "if [he] had to use this particular program [he] would have searched the manual for instructions."

In Table 4.2, we show a summary of our tagging for all the sessions we observed. The presence of "*" at the intersection between the column with the participant's ID code and the line with a specific tag indicates the presence of that tag in the analysis of the participant's interaction with Audacity. Bearing in mind the set of 13 tags presented in Chapter 3, their definitions and (sub-) categorization (pp. 43–46), we see that only 10 of them were used. "Help!," "What is this?" and "Thanks but no, thanks." do not figure on Table 4.2. Regarding "Thanks but no, thanks.", we found no evidence whatsoever of the symptoms associated to that tag. Regarding the other two tags, we did find occasional symptoms of "What's this?," like users reading tool tips, but interpreted them as epistemic interactive steps that are natural and necessary in first encounters with systems. In our view these were actually successful cases of metacommunication, helping first-time users resolve expected breakdowns. The single instance of "Help!" was when E2 turned to the observer and asked for just-in-time information about Audacity's feedback. But, because this occurred in the middle of a major breakdown tagged as "Why doesn't it?," we thought this piece of evidence did not add to the bigger communicability problem that was clearly going on. It is nevertheless noteworthy that only one participant (E3) read the online manual. The other novice users of Audacity, even if caught in major interaction breakdowns, did not resort to the online manual, trying to guess the solution instead of "asking for it" explicitly [cf. the paradox of the active user (Carroll & Rosson, 1987)].

The only two participants that failed completely in producing their ringtones, N2 and E2, are the only ones whose interaction has the '*I give up.*' tag. Interactions of participants that failed to export the MP3 file in the end, but thought they had done it (N1 and N3), were tagged with '*Looks fine to me.*' Notice that we used '*Looks fine to me.*' for E2, as well. This is because in the end of the session E2 *saved* his ringtone (using 'save as', not 'export') with the name 'filemp3' (no extension). In the posttest interview he confirmed that *he had* saved the file in MP3 format (although he hadn't).

Table 4.2 also shows that some tags appeared more frequently than others. This is the case of "Why doesn't it?" and "Where am I?". The number of "*" indicates only the presence (but not the intensity or significance) of the corresponding breakdown. For example, N1, N2, N3, E1 and E2 experienced problems tagged as "Why doesn't it?" and "Where am I?," but we do not see how long the breakdowns lasted, or how many times they recurred during a given participant's interaction with Audacity. By counting the recurrences of the same pattern of breakdown for each participant (e.g., two recurrences of "What now?" for N3, and three recurrences of "Why doesn't it?" for E1), the most recurring tag was "Why doesn't it?" (a total of eight recurrences in five out of seven tests). The second most recurring tag was "What now?" (six, in three tests). The third most recurring tags were "Where am I?" (five, in five tests) and "Looks fine to me." (five, in four tests).

Tags with higher distribution indicators—"Where am I?" and "Why doesn't it?" (both tagged in five of the seven tests)—unveiled important interactive issues. Most, although not all, of the recurrences of "Why doesn't it?" correspond to the participant's attempt to select audio by manipulating the timeline, instead of track contents (see Figure 4.12). The symptom of the breakdown was the participant's persistence in trying to achieve selection with manipulations that were clearly *not* helping him or her achieve the task. An interesting aspect of this particular situation with Audacity is that users are *doing the right thing in the wrong place*. The selection of the audio span is correct, except that when done on the timeline it means "*play* this selection," and when done on tracks it means "*take* this selection." Hence, each of the "Why doesn't it?" taggings corresponding to this special kind of interaction was simultaneously tagged with a "Where am I?." In other words, a failure to understand what the designer's deputy is saying and subsequent attempts to clarify this communication by repetition (or autonomous sense-making, according to Table 3.1) was actually associated with missing how different contexts of manipulation are signified in Audacity.

Our specific use of "Where am I?" in this situation is *nonstandard* according to our own subcategorization criteria presented in Table 3.1. The *standard* criteria for using "Where am I?" is that

Categorization	Distinctive Feature	Tag
Complete Failures	User is conscious of failure.	"I give up."
	User is unconscious of failure.	"Looks fine to me."
Partial Faliures	~~User understands the design solution.~~	~~"Thanks, but no, thanks."~~
	User does not understand the design solution.	"I can do otherwise."
TEMPORARY FAILURE **1. User's semiosis is temporarily halted**	because he cannot find the appropriate expression for his illocution	"Where is it?"
	because he does not perceive of understand the designer's deputy's illocution	"What happened?"
	because he cannot find an appropriate intent for illocution	"What now?"
TEMPORARY FAILURE **2. User realizes his illocution is wrong**	because it is uttered in the wrong context	"Where am I?"
	because the expression in illocution is wrong	"Oops!"
	because a many-step conversation has not caused the desired effects	"I can't do it this way."
TEMPORARY FAILURE **3. User seeks to clarify the designer's deputy's illocution**	~~through implicit metacommunication~~	~~"What's this?"~~
	~~through explicit metacommunication~~	~~"Help!"~~
	through autonomous sense-making	"Why doesn't it?"

More occurrences ■ ■ ■ ■ □ Less occurrences

FIGURE 4.17: Most frequently occurring tags in CEM tests.

"user realizes his intended interaction is wrong," but in CEM tests, the extent of this realization is debatable. Of course the users knew that *something* with their selection was wrong (and this is why they repeated the selection insistently, trying to find out what it was). But *they* did not necessarily know that the problem was with "the wrong context." In posttest interviews there is no evidence that in this case they were thinking about *context* (nor that they were not). However, for the external observer, who knows how to interact with Audacity, it is absolutely clear that, as mentioned above, they were *doing the right thing in the wrong place*. This is why we decided to register this breakdown as a case of "Where am I?" in association with "Why doesn't it?."

Figure 4.17 summarizes important qualitative aspects of our tagging and interpretation, especially in view of the purposes of triangulation, discussed in the last section of this chapter. The most frequently tagged breakdown was "Why doesn't it?," which gains especial importance in contrast with the other two tags of the same category, "What's this?" and "Help!" On the one hand, as already noted, the occasional reading of tool tips (which when constant or extensive is indicative of a problem with the meaning of static signs in the interface), was not interpreted as a breakdown. But on the other hand, we had so many occurrences of "Why doesn't it?," showing that metacommunication through tool tips and online help (access to the manual and wiki) and use of metalinguistic signs in Audacity was ineffective.

The second most frequent tag, "What now?," is also indicative of a severe breakdown. Compared to the other breakdowns in the same category ("Where is it?" and "What happened?"), "What now?" indicates a situation where the user cannot even formulate an intention to be communicated to the system. For example, the presence of "What happened?" may or may not indicate a severe metacommunication problem. It is not so severe if the user's failing to perceive or understand a specific message that the system is sending leads to a breakdown that is quickly resolved by an implicit request for clarification ("What's this?") or trial-and-error ("Oops!"). However, "What happened?" may represent a severe breakdown if it takes the user a long time to resolve the problem, or if the problem develops into a situation where the user is lost. In our tests, we had a single occurrence of "What happened?" associated with zoom. The user pasted a span of audio from one track into a new file and the system automatically zoomed the new track to "fit to window view." As a result, the user lost the visual cue that identified the pasted audio span as the *same* as he had copied. In spite of this single occurrence, because "What happened?" pointed to such an important communicability problem (loss of visual identity of objects in direct manipulation interfaces), the category of temporary failures associated with disruptions in the users' semiosis stands out as a particularly relevant one in the communicability evaluation of Audacity.

We would finally like to remark that although our tagging is concentrated in the category of *temporary failures*, from which users typically recover, and although only one of six first-time users (N2) gave up interacting before doing at least one of the main test steps, this does not necessarily attenuate the problems with metacommunication in Audacity. In posttest interviews, most participants gave us evidence that they missed important elements of the designers' message. This is true not only for novices but also for participants with digital audio-editing experience. Notice that even AU, an experienced Audacity user, missed one of the nice features of the system. Here are some excerpts of their interviews:

> I could not understand how it works. I was just trying to guess. [. . .] It's not for me. (N1)
>
> I found it complicated. [. . .] When I did something it was just by lucky chance. (N3)
>
> I found out how things worked completely by chance. (E1)
>
> Selection is very confusing. There are many options for selecting, and I could not understand the difference. (N1)

> Selection is really very confusing. (E1) This participant worked with the "select region" menu during all the editing, which is a more complicated way to interact with Audacity than simply selecting portions of the timeline directly. This suggests that he did not quite understand how selection works.
>
> I do not think all tracks were rendered together . . . or maybe they were. Maybe this is what happened, yeah . . . But it's not what I figured was happening [during the test]. (E3)
>
> I know [timeline selections] are there, but I did not know I could do this [using timeline selections to navigate quickly to different playback portions]. (AU)

Additionally, when talking about his experience, E2 said that he *had understood* how selection works, but gave us the wrong explanation. Then, he stopped for a while (as if listening to what the had just said) and corrected the explanation. We took this as an instance of "after the fact" reasoning, suggesting that his successful selections might also have been, at least partially, a matter of *chance*.

In this particular case study, we have explored three of the four perspectives proposed to guide interpretations in CEM: the frequency and context of occurrence of each type of tag, the existence of patterned sequences of tag types, and the communicability issues that have caused the observed breakdowns. The level of problems signaled by the occurrence of tag types and sequences was not relevant in the context of our research question.

4.2.3 Semiotic Profiling

In this last stage of CEM, we return to the purpose of applying this method in our study (triangulating SIM results) and to our research question, stated in the beginning of the chapter: which strategies are chosen to communicate a system's basic functions and how they can or should be used?

The empirical data obtained with CEM suggests that the designers' strategy to rely heavily on metalinguistic signs in Audacity's online manual and Wiki is ineffective in preventing severe communicability problems and letting the users know all of the advantages of using Audacity. All semiotic resources in the interface (metalinguistic, static, and dynamic signs) have the power to stimulate the user's semiosis in the direction of productive hypotheses. The quickly assembled ad hoc theories about how the system works (e.g., Carroll & Rosson, 1987) are deeply influenced by such resources and must either be the correct ones, or have the power to bring the user's interpretation out of stagnant semiosis stages and into conversations that can eventually lead to the right one. This is an important difference between cognitive evaluation criteria and semiotic ones. For example, whereas from a cognitive engineering perspective (Norman, 1986) moving from one mistake to another does not help the user, from a semiotic engineering perspective this may represent a valuable opportunity to *reconnect* with the designers' interactive discourse, and thus restore conversation. Designing redundant messages that can increase the chances to restore and reconnect metacommunication in breakdown situations is thus an important strategy that Audacity's developers might adopt to improve the system's communicability.

The above conclusion is reinforced by the fact that test participants did not understand some basic principles in Audacity's interface, as posttest interviews showed. Some of them explicitly said that success was a matter of *chance* rather than *comprehension*. Others clearly said (or showed) that they were confused by interface signs. Therefore we are convinced that communication achieved by static and dynamic signs does not convey the message expressed with metalinguistic signs. An important question to ask is whether they are communicated by the same speakers. Although from a user's point of view the answer should probably be obvious—"Yes, Audacity is speaking!"—at a closer look, reading their own material online, we may suspect (or realize) that "No, there are many people speaking (volunteer developers)." The latter is a valuable clue to the origins of metacommunication problems with Audacity, which we will discuss when we report the conclusions of the whole study, shortly.

User observation in CEM gives us a privileged insight into the interpretation of dynamic signs. During test sessions, we can trace how the system's behavior influences the decisions of participants in the context of their current activity. So, whereas our results with SIM pointed to certain kinds of problems, CEM can tell us more.

Regarding the problem with interpreting the system's states, especially for novices, CEM clearly showed that using "pause" and "stop" controls correctly was a problem. Users quickly understood that in order to edit audio they had to stop the playing. As one participant explicitly remarked in the posttest interview, "the disabled menu functions annoyed me; there are too many buttons to use for play, stop, pause—it should be simpler" (N1).

The most interesting results with CEM, however, were that communicability evaluation revealed much deeper issues involving visual and auditory representations of audio than we could capture in our semiotic inspection.

In Audacity, there are three very salient representations of audio: sound, timeline, and waveforms. Each plays a different function in the interface: sound *identifies* the object of interaction reproducing audio *as it is* in the current state of the system; the timeline places the focus of interaction on the audio extension over time and enables interactive discourse related to temporal dimensions (e.g., "play/select audio comprised between 00:01:34 and 00:02:01," or "name audio comprised between 00:01:34 and 00:02:01 as *encore*," etc.); and finally waveforms enable interactive discourse focused on *atemporal* dimensions of audio, such as waveform patterns, pitch and spectrum (e.g., "play/select [this] quieter part," "lower this sample," "remove hiss from this passage," etc.). We will refer to waveform representations as *spatial*, because in order to understand them users must interpret the 2D structure of track visualizations correctly.

The designers' choice of direct manipulation, considering that menus do not support all of the editing functions in Audacity, requires that users not only interpret temporal and spatial representations of audio correctly, but that they know which dimension of *undivided sound* coming out of the speakers they must manipulate in order to achieve a particular goal that they have in mind. CEM tests have shown that all novice and at least one experienced first-time user (E1) were very confused by metacommunication associated with temporal and spatial representations. Additionally, E2 and E3 were less confused than others, but still experienced some revealing breakdowns. E3, like most other participants, was confused by the meaning of parallel tracks that automatically appear when the user records something with the microphone—are they rendered together when you export MP3 files or not? Also, N3, E1, and E2 had problems with zoomed visualizations—momentarily, they were unable to tell which part of the audio was playing (because when zoomed in, they lost the temporal and spatial context of the *whole* track), which confirms one of our results with SIM.

4.3 FINDINGS AND CONCLUSIONS

After having used SIM and CEM to evaluate Audacity, we were able to contact three members of the Audacity Team by email. They have different roles in the design and development of the system. We presented the main ideas involved in semiotic engineering, SIM and CEM, and the goals of our research. Then, without telling our results, we sent them a questionnaire, which they agreed to answer. First, we asked how they would instantiate the first half of the metacommunication template: "Here is my [the designer's] understanding of who you [users] are, what I've learned you want or need to do, in which preferred ways, and why." Then we asked what was difficult to do in Audacity and why, and what they thought should be improved and why. We also asked them what were the best HCI design solutions in Audacity 1.3.5 compared to previous versions and/or other audio editors, and what was particularly good about them. Finally, we presented and discussed our findings with them.

Regarding the instantiation of the first half of the metacommunication template, they said that Audacity is for "a very wide range of users" and listed many examples of user profiles. However, it was clear from their answers that each one had greater interest in one or two specific user profiles. So, as two of them had explicitly anticipated to us, the metacommunication template was instantiated in considerably different ways.

One of them added an important element to validate our research results:

> Typically a developer will implement a new feature, and there's a wide range of knowledge of HCI principles among the developers (e.g., several have never heard of Fitts Law, etc.), so these new features have a wide range of usability.

There are two important aspects to emphasize based on what they said. One is a confirmation that the problems with fragmented or inconsistent metacommunication most probably originate in the development process, which is not strictly planned and structured. Each volunteer has specific motivations, and implements new features according to their own interests and abilities. As a result, the development process is somewhat haphazard, something that occasionally surfaces in the form of interface and interaction inconsistencies.

The other important aspect to remark is that there is a wide range of HCI knowledge among team members and that the token of expertise spontaneously offered by one of the respondents is Fitts Law. This is one of the best examples of how a scientific theory can be applied, in practice, to leverage the quality of the users' experience with various types of input devices such as a mouse, a digital pen, a stylus, and so on. HCI research has not produced, however, similarly popular and widely accepted scientific theories that can be applied to solve higher-order and more abstract problems in interaction design, such as the ones detected in our case study. Notice that as we move from psychomotor to semiotic aspects of human behavior involved in HCI design, we may also have to move from *predictive* to *non-predictive* foundational theories required for the development of tools and techniques to support HCI design.

The three Audacity Team members that answered our questionnaire expressed their conviction that direct manipulation was the right choice of interface style. One of them said that "menus are long and lack visual guidance." Another one mentioned that "most users (. . .) prefer mouse and direct manipulation, then menus and shortcuts as they become more expert." All of them seem to believe that direct manipulation is easier for novices as a rule. However, direct manipulation is easier for novices only if the interpretation of visual metaphors and how they can be used to communicate with the system is obvious to the users. Our analysis of the semiotic challenges involved in interpreting and using visual representations of audio has shown that this is not the case in Audacity. On the contrary, we have strong indications to suspect that if there were simple menu options for such basic things as "record" or "save as {Audacity Project, Exportable Audio Format}" our participants would have had much less difficulty to run the test scenario successfully.

After we presented our results, the three Audacity Team members engaged in a stimulating *follow-up* discussion that contributed to validating our findings. Moreover, the discussion gave us evidence of the epistemic value of our results for our collaborators. Their attitude toward our results was remarkably positive, which does not mean that they agreed with all of ours results. In general, they expressed enthusiasm but also surprise and disagreement. One member of the team said: "The results are very enlightening." Notice the use of the word *enlightening*, suggesting that our results were not taken as technical guidelines but as actual new *knowledge* that shed new light on certain aspects of the Audacity interface design. Another emphasizes a broader perspective stimulated by our results and suggestions: "Fascinating. This looks like an awful lot of preliminary work to me, and it clearly shows some majors problems that users have with the software." Notice that copresence of terms "preliminary work" (meaning there is much more to be known and done) and "clearly shows some major problems that users have" (expressing the perception of recurring categories of design issues with Audacity). Both comments suggest that semiotic engineering methods, in this study, have clearly contributed to what Schön (1983) refers to as *an epistemology of practice*, helping developers reframe problems and/or reconsider solutions.

Also, because of the epistemic character of our results, Audacity Team members brought us what one of them referred to as "questions and criticisms" regarding some of our findings. One example is related to their conviction that direct manipulation is the best choice of interface style and that users do not like to use menus. Two of Audacity members were very surprised with the problems users had with menu choices and structure. They mentioned that in their experience users frequently are impatient when interacting with menus.

In follow-up discussions, after the interview and presentation of results, other kinds of statements emerged that help us validate our findings. One of them is related to the difficulties involved in metacommunication made by a team of volunteers following free and open-source systems development practices. In such conversations, the three members of the Audacity Team mentioned the challenges present in FOSS development. They said that frequently they have different viewpoints or preferences and often these differences are simultaneously communicated thorough the interface. In particular, one of them told us that he misses HCI tools that can help the team produce a consistent and coherent discourse, either at design or implementation time. With interest, he asked:

> How do you see that in the context of an ongoing open-source project where various people dip in and out at will? (Which of course is the context of this study.) That seems to be another problem which perhaps you haven't addressed?

As we will discuss below, we have addressed this problem as a challenge to be faced by semiotic engineering researchers.

The other important aspect questioned by two of our collaborators is related to what we called a shift from *predictive* to *non-predictive* foundational theories. This shift is often required for the development of tools and techniques to support HCI design. In previous chapters, we said that theories and methods that can be applied to solve higher-order and more abstract problems in interaction design are probably non-predictive and qualitative. However, the qualitative paradigm still has its problems in HCI research and professional practice. Two Audacity team members questioned various methodological aspects of our qualitative approach. For example, they asked why we worked with such a small sample, what validation criteria we used, etc. Again this discussion and questioning provided strong evidence of the reflective and epistemic nature of the semiotic engineering contributions in the technical contexts of professional practice.

However, we must not forget that our purpose goes *beyond* Audacity, into the HCI domain itself. If we take a second look at the contributions of SIM and CEM for this study, we will see that our results point in directions that actually transcend the scientific investigation about the designed strategies to communicate a system's basic functions and how they can or should be used. They allow us to:

- Identify more general HCI problems (i.e., not specific to Audacity) that semiotic engineering probably can solve with further developments of the theory (internal articulation);
- Identify more general HCI problems that semiotic engineering might be able to solve with input from other sub-areas in computer science (cross-articulation); and finally
- Identify issues related to other sub-areas of computer science that semiotic engineering might contribute to advance (external articulation).

4.3.1 Internal Articulation

The main results of this study suggest that strategies for communicating basic functions are critical for *all* users—precisely because all need to use these functions. Organizing efficient and effective metacommunication addressed to *all users* is not a matter of showing more or less components to different users, but rather one of articulating specialized discourse about the same theme. Users with different profiles should have different kinds of conversation with the system about the same set of basic functions. By exploring redundancies between metalinguistic, static, and dynamic signs, we can *engineer* metacommunication that considerably increases the chances that the users' semiosis will include elements of the designers' semiosis encoded in interactive systems. Some redundancies are direct, saying the same thing with different signs; others are indirect, saying related things that will point in the same direction.

A crucial aspect in communicating with users at design time is that the development team be capable of *producing* a unified, cohesive and coherent discourse (Barbosa & Paula, 2003; Paula, 2007). Using the metacommunication template as a design artifact to represent collective discourse seems to be a simple but promising idea to improve software communicability. In concrete terms, the metacommunication template could be used to structure computed-mediated communication in online forums and discussion lists (Barcellini et al., 2007; Nichols & Twidale, 2003; Twidale & Nichols, 2005).

Other semiotic engineering theorists have proposed a language to model the conversations in which the designer's deputy can engage at interaction time—the Modeling Language for Interaction as Conversation (MoLIC; Barbosa & Paula, 2003; Paula, 2007). In Figure 4.18, we show

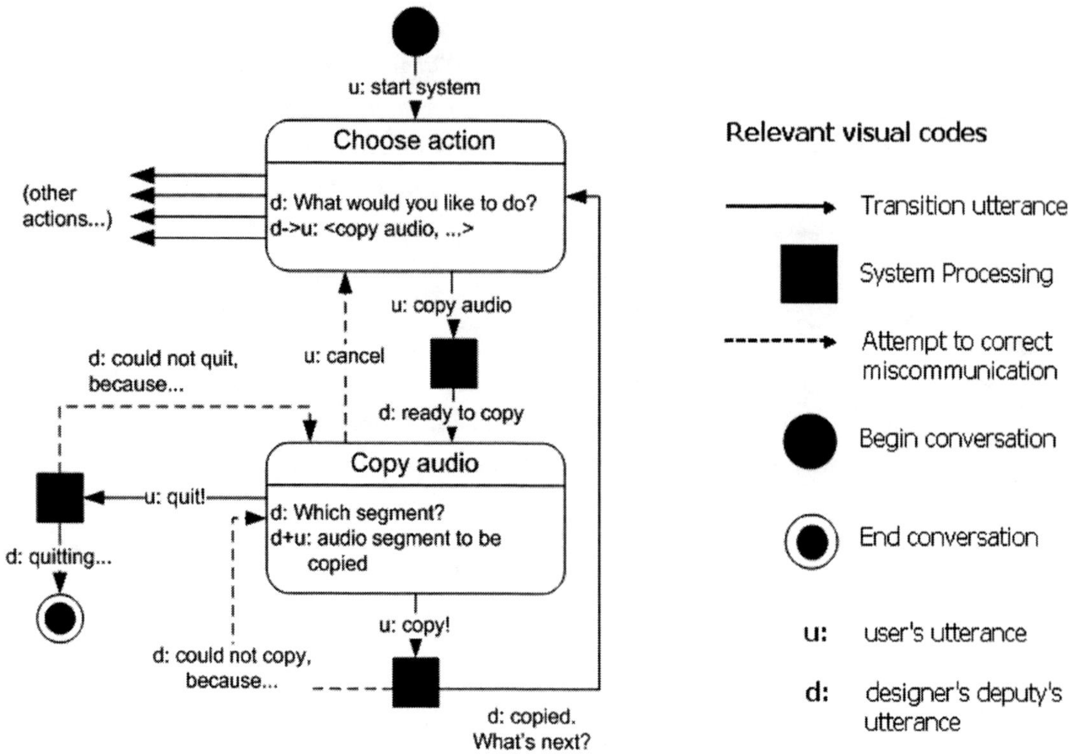

FIGURE 4.18: A simple MoLIC diagram.

what a MoLIC diagram looks like. Larger rounded boxes represent conversational *scenes*, dialogs between the user ("u") and the designer's deputy ("d") about a single topic (e.g., "Copy Audio"). The epistemic use of MoLIC diagrams should help designers express the *logic* of conversations that the designer's deputy can have with users. At this level of abstraction the selection of interface signs (metalinguistic, static, and dynamic) is not fully specified, and the correction and completeness of specifications is secondary to the possibility that a MoLIC diagram will "talk back" to the designers (Schön & Bennett, 1996), and thus give them the opportunity to discuss their views and build or strengthen common ground. The diagram helps them see larger contextualized portions of metacommunication and identify recurring patterns of communication, such as the enabling conditions for "Cancel" and "Quit" messages, the content and context of error messages, etc.

Additionally, if designers begin to produce such representations, a new empirical evaluation method that captures not only the communicative breakdowns (as CEM does) but also the positive evidence of mutual understanding, is instantly in place. The evaluator can map user interactions onto MoLIC diagrams and see how close the actual user–system conversations follow the design model. The difficulty with MoLIC as we write, however, is that it is still a work in progress, with relatively little critical mass to enable a more extensive use of it in software development activities.

4.3.2 Cross-Articulation

Contemporary wide-spectrum applications such as *media* editors (including text, images, sound, etc.) have widely diverse users, which tend to have very specialized views and uses of such systems. Thus, designing appropriate interaction and interfaces in these cases is a major challenge. Sophisticated mechanisms have been proposed to allow end users to extend, customize and adapt software to their particular needs (Lieberman, Paternò, & Wulf, 2006; Silva & Barbosa, 2007). Nevertheless, as one of Audacity's team members said in his interview, "there is perceived reluctance of less experienced users to go into Preferences to change settings, while more advanced features drives a need for more preferences (which makes them more intimidating again)."

In order to improve this situation, as mentioned in the conclusions of SIM, we might examine research in AI and see if we can combine dialog and user models proposed for intelligent conversational systems (Moore, 1995) with semiotic engineering interaction models (represented as enhanced and annotated MoLIC diagrams). This would allow us to generate *interaction templates* that users might be able to choose from preference dialogs like the one shown in Figure 4.16. Notice that this alternative is not the same as a composition of modules shown or hidden in the system's interface. It involves the elaboration and activation of different discourse strategies, depending on communication preference parameters that users may select implicitly or explicitly at interaction time.

Another important issue raised in the conclusions of CEM—the direct manipulation visual representations of temporal and spatial dimensions of objects perceptually and cognitively interpreted as a unity (like sound, image, speech)—also gives us an opportunity for cross-articulation. For example, the behavior of structured representations in computational space and time, which Gelernter and Jagannathan (1990) have coined as *Ideal Software Machines*, has been defined as a theoretical base model for analyzing programming languages, in general. By mapping manipulations of spatial and temporal interface objects onto the structure of ideal software machines, we can characterize direct manipulation interfaces as a particular case of visual programming and begin to think of interaction as programming. Among the potential advantages of this perspective is the possibility of promoting greater *computing* literacy among users, which goes beyond *computer* literacy. Inasmuch as more users begin to view computer programs as a rule-governed play on representations, given the appropriate interfaces they might gently transition into small-scale programming, which would be beneficial not only for them but also for software developers (Lieberman, Paternò, & Wulf, 2006).

Another opportunity of cross-articulation begins with deeper semiotic analyses of sign types used in visual representations used in direct manipulation interfaces. In a past study (de Souza &

Sedig, 2001), we used Peircean classification schemas to analyze the visual interface of educational software. The results, as we now see, hold the promise of informing the choice of metacommunication strategies for systems like Audacity and many other media editors. Very briefly, in that study we concluded that visual representations evoking mainly or exclusively the similarities between the representation and the perceptual characteristics of the represented object (like its shape, color, texture, etc.) work best when direct manipulation affects the object as a whole, or only and exclusively the evoked characteristics of the object. Direct manipulation of separable dimensions, factors, parameters, and features of the object, however, may get very problematic if visual representations evoke something other than the abstract facets of the object that are affected by computations.

4.3.3 External Articulation

The semiotic engineering study with Audacity also indicates very clearly a new research topic that the theory has only dealt with in passing up to now (de Souza, 2005). We will now refer to it as *recursive metacommunication structure*, that is, how metacommunication in the lower levels of software structure, such as frameworks and programming languages, for example, affects metacommunication at the top level of the structure—the user interface.

FOSS development practices, as well as commercial development, extensively explore the use of components. An interesting post in Audacity's public forum (November 10, 2008, http://audacityteam.org/forum/viewtopic.php?f=21&t=7167) illustrates communicability issues with software components that semiotic engineering should examine. The message asks:

> Is there any way to invoke Audacity via a VB Script? If so—can I utilize parameters to provide information to Audacity? What I would like to do is to automate the file naming and MP3 conversion after a recording. In other words—when we are done recording, the operator simply exits Audacity and the conversion takes place. If this is not possible—other suggestions are appreciated. I need to eliminate some of the manual steps in the mp3 conversion—too many mistakes are happening.

In addition to restating one of our results with CEM, the posted message shows that the developer's idea is to build a VB Script interface to one of Audacity's basic functions, "export." According to http://audacity.sourceforge.net/community/, the programming language used in Audacity's development is C++, a general-purpose object-oriented language based on C. VB Script, however, is primarily a web-scripting language, designed to be used in conjunction with HTML. This difference in conception affects the kinds of programs that each language is apt to encode, and consequently the users' experience with such programs. So, when developers combine one with the other, we can expect to see the traces of change in software encoding techniques reflected in static and dynamic signs that compose the user interface. Will the combination be able to preserve cohesion and coherence in metacommunication? Under which conditions and to what extent? What are the consequences for users?

The current answer to the above questions is: we, semiotic engineering researchers, don't know. Therefore, developers do not have the necessary knowledge support to inform their decisions in situations such as the one expressed in the message above. What we do know is that static and dynamic interface signs can reveal communication breakdowns experienced by developers.

As we were writing this book, one of our students, who is very familiar with semiotic engineering and also an experienced software developer, gave us one of the most striking spontaneous evidence of the potential depth of metacommunication issues in software development. As part of his course projects, this young man decided to use SIM and CEM to evaluate a graphic editor that he developed using the Eclipse platform.[3] The special interest of his projects was that he was evaluating *his own* design and development. In both SIM and CEM evaluations, he was surprised to discover that he had problems with metacommunication in Eclipse. He *meant* to send his users one kind of message, but SIM and CEM revealed that the Eclipse components that he selected to express his intent "betrayed" him in important ways. In one case, the component's behavior had communicative side effects that he wished to avoid because they were likely to cause interactive breakdowns for the users. He struggled with the platform, but did not find a way to *silence* metacommunication from *others* speaking through *his* interface. In the other case, he observed a user expressing himself with a combination of static and dynamic signs that our student expected not to work together. But they worked perfectly well. Observing this user's interpretation of the interface, he realized for the second time that there are *hidden meanings* in Eclipse components, of which the developer has very little control (if any). He was not only puzzled by the recurring evidence but also challenged by the fact that component documentation typically does not address the kinds of issues that, as he realized, may so deeply affect interaction design.

Inspecting the recursive metacommunication structure thus places developers in the position of users and raises communicability issues originating in the deeper layers of software design and development. The new perspective advanced here is that of *programming as interaction*, the dual of *interaction as programming* advanced in our discussion of cross-articulations above. We take this two-way connection between programming and interaction as evidence of a strongly cohesive relationship that semiotic engineering could and should investigate scientifically. The relevance of the investigation is to show that this particular theory of HCI also has the potential to contribute with new knowledge for core areas of computer science such as programming and software engineering.

* * * *

[3] http://www.eclipse.org/org/.

CHAPTER 5

Lessons Learned with Semiotic Engineering Methods

In this chapter, we analyze the main lessons learned with SIM and CEM. First, we examine *why these two methods have the potential to generate new knowledge*, taking into consideration their limitations and specificity. Then, based on our case study results, we examine *what kinds of new knowledge can be achieved* by their application.

We believe that the potential of a method to achieve new knowledge is closely related to the research paradigm where it belongs. Semiotic engineering methods follow a nonpredictive paradigm in research. Given the foundations of semiotic engineering presented in Chapter 2, *meaning* is a continuous and unlimited interpretive process, rather than an ultimate value (Peirce, 1992–1998; Santaella, 2004). This perspective is incompatible with that implicitly or explicitly assumed by predictive methods:

> There is no room for a pure objective and stable account of meaning (. . .) Meaning carries inherent subjective and evolutionary ingredients determined by unlimited semiosis that cast a shadow of doubt upon the idea that the users' context, requirements, capacities can be fully captured by any human interpreter at any given time. (. . .) Consequently, in this sort of epistemological context a researcher cannot possibly assume a positivist attitude, commonly adopted by many who aim to build predictive theories. From a semiotic perspective one cannot observe and interpret empirical phenomena without being affected by his or her own subjectivity, and the sociocultural context around him or her. Therefore, the value of observation and interpretation cannot be dissociated from the purposes the empirical study is designed to serve in the first place (de Souza, 2005, p. 28).

SIM and CEM thus subscribe to a nonpredictive paradigm in research, in which *qualitative methods* are used to explore the evolution of meaning, rather than ideal and definitive meaning configurations (Cresswell, 2009; Denzin & Lincoln, 2000). Both methods are tools to support systematic analysis and interpretation of *singularities*. When using them, researchers actively position themselves in the context under investigation, exploring it without previous hypotheses in search of meanings that will expand previous knowledge. It is thus clear that SIM and CEM do not follow experimental and/or predictive scientific traditions that aim to measure and/or control variables in empirical phenomena. As a consequence, like other qualitative methods, SIM and CEM create the possibility for finding the unexpected and the new, following the trail of new meanings that researchers would not have been able to anticipate.

Other than semiotic engineering, there are a number of HCI approaches, frameworks, or theories that promote an in-depth, exploratory, situated, and interpretative perspective on research and naturally adopt qualitative methods of investigation to achieve their goals. For example, the works of Nardi (1996), Suchman (1987), and Dourish (2001) use etnomethodology to explore, interpret, and understand unknown and unpredictable situations involving users and computer technology. This, as noted by Shneiderman (2007), is a promising feature for innovation.

However, the distinctive feature of semiotic engineering methods as compared to ethnography, for instance, is their strong theoretical commitment. Considering the research strategies defined by Cresswell (2009), SIM and CEM fit into the transformative research strategy. From Cresswell's perspective, transformative research explicitly brings theories to their inquiries as "an orienting lens that shapes the type of questions asked, who participates in the study, how data will be collected, and the implications made from the study" (p. 208). This kind of approach leads to "an overarching perspective" (p. 208) that is more important in guiding the study than the use of the methods themselves. Thus, the theory shapes a directional research question, creates sensitivity to collecting data, and ends with a call for action and change.

In the context of semiotic engineering research, as already discussed, because SIM and CEM are based on the ontology of semiotic engineering, they are meant to support scientific investigations exclusively centered on designer-to-user *metacommunication*. Metacommunication gives direction and limits to our methods. In other words, SIM and CEM can only help us gain knowledge about the designers' communications computationally encoded in the form of interactive messages presented in a computer system's interface. Consequently, SIM and CEM cannot support the analysis of *all* the potential for communication between designers and users (e.g., all that designers would wish to communicate or think they are actually communicating to the users of interactive systems they design). The methods are only prepared to analyze communicability within the limits of the designer's deputy interactive discourse.

As researchers explore the emission and reception of the designer's deputy's interactive discourse, they must carefully examine how SIM and CEM map the territory of their investigation, paying attention to the limits and boundaries imposed by each method. On the one hand, they must look inside these limits and boundaries (the internal perspective) and decide whether SIM and CEM can be used to help them find appropriate answers to the research question they propose to explore. On the other hand, they must look outside these limits and boundaries (the external perspective), asking themselves how SIM and CEM can or should be complemented by other HCI methods or even by methods used in other areas of scientific research.

Taking an internal perspective, it is possible to see that SIM and CEM can be used to analyze strategies (e.g., using context-sensitive menus) with which designers communicate that they have followed a particular usability principle (e.g., "recognition rather than recall"). This perspective also shows that SIM and CEM procedures are inefficient, if not ineffective, if the research question aims to detect or discover new usability principles. By the same token, an investigation of the user's performance and productivity, or the user's learning curve, falls outside the scope of semiotic engineering methods. In sum, by adopting an internal perspective of analysis, we see that the only research questions that fall *within* the territory mapped by SIM and CEM are those that can benefit from an in-depth exploration of metacommunication achieved through the designer's deputy's discourse.

Although the limits exposed by an internal perspective examination show that the scope of SIM and CEM application is rather narrow, they also show that these methods are considerably permeable. The investigation of metacommunication in HCI is a highly abstract goal. It can be instantiated for a wide range of domains, contexts, and purposes. So these methods can, in fact, string together a number of very diverse classes of interactive systems, allowing us to compare them with each other in terms of some specific aspect(s) of metacommunication. This is an interesting possibility in a research territory that has been diagnosed with problems of "scientific fragmentation" (Carroll, 2003, p. 5). To illustrate this point, we can think of two widely different types of systems—safety-critical control systems and computer games. The former have an imperative need to avoid interactive errors, if not to eliminate them completely, because the cost of errors may, for example, be paid with human lives. The latter, however, stimulate errors, challenging users to outperform the system and win the game. In spite of this clear opposition, both kinds of systems involve strategies to communicate errors. How do they do it? Which communicative strategies can be used in one case or the other? Do they have anything other in common than their referent object type (i.e., errors)? Do strategies that communicate opposing message contents need to be opposing strategies? Researchers interested in identifying and analyzing communicative strategies in such disparate domains as safety-critical systems and computer games can use the methods proposed by semiotic engineering as useful epistemic resources that will help them relate concepts and phenomena from two sub-areas of interest in HCI that apparently have very little in common.

In an external perspective, there are two questions to be asked. One is whether SIM and CEM can benefit from input produced by other methods. The other is whether SIM and CEM can themselves produce input to other methods. As mentioned in Chapter 2, the object of investigation that both methods explore is the designer's deputy's interactive discourse, which is, in essence, a computational object. Therefore, SIM and CEM require that this object be specified in sufficient detail, allowing for at least partial identification of the fundamental components involved in the process: senders, receivers, communicative intent, content, and expression.

We thus see that SIM and CEM *cannot* (and are not meant to) support the collection and analysis of preliminary data to compose the metacommunication message. Consequently, semiotic engineering research can greatly benefit from results of ethnographic methods to examine how the content and effects of metacommunication relate to the broader context of reception. More specifically, ethnographic methods can expand the depth of analysis referring to the first segment of the metacommunication template: "Here is my understanding of who you are, what I've learned you want or need to do."

At the other end of the spectrum, SIM and CEM can provide interesting input for the analysis of *expressionistic* computer artifacts, that is, computer systems that are built *without* a systematic concern with users. For example, the emphasis may be on the affective states, ideas, and talent of the *author* of the program or system, as is the case with computer art (Bentkowksa-Kafel, Cashen, & Gardiner, 2008). Using SIM and CEM to explore how such systems are received may add interesting dimensions to the analysis carried out by other methods (e.g., semiotic methods, aesthetic analysis, etc.). Are there underlying patterns of communication to be revealed? Do static, dynamic, and metalinguistic sign combinations signify higher-order meanings? These questions and others can also contribute to research in other sub-areas of computer science such as multimedia systems, somputer graphics, and games.

Other limitations of semiotic engineering methods refer to supporting research about the impact of computer technologies *beyond interaction*. SIM and CEM have not been conceived to support longitudinal studies. The designers of a given technology can communicate effectively and efficiently their design vision and principles, yet the technology may not be adopted by the targeted users. The exploration of factors influencing technology adoption is clearly outside the scope of both methods. Again, ethnographic methods can be productively used in this kind of research context. In this respect, it is noteworthy that semiotic engineering can be articulated with ethnography at both ends of a *continuum*—ethnography can produce input for the design of metacommunication and also explore the long-term effects of metacommunication in broader social contexts.

Additionally, when discussing how SIM and CEM can complement or be complemented by other methods, we must examine consistency between scientific paradigms. Because semiotic engineering methods follow a nonpredictive paradigm in research, they can only be articulated with other methods that follow the same paradigm. Therefore, SIM and CEM are incompatible with methods that work with hypothesis refutation and aim to support generalizations.

Finally, as already discussed in Chapter 3, validation of SIM and CEM results is achieved by triangulating results between different nonpredictive theories and research results. Traditionally, predictive scientific theories have been constructed using *quantitative* research methods. However, recent discussions in the field of scientific methodology have pointed out that under special circumstances qualitative and quantitative methods can be used to complement each other (Cresswell, 2009; Denzin & Lincoln, 2000; Patton, 2001). The possibility to combine SIM and CEM with quantitative methods for triangulation purposes is, at the moment, an open question in semiotic engineering. We have only begun to experiment with this kind of combination, and do not have solid results to discuss yet.

In Chapter 3 (pp. 23–24), we said that scientific methods lead to new knowledge by producing: a new account of known problems, using theoretical concepts that support the formulation of relevant research questions; the identification of new solutions, partial or complete, generic or specific, to known problems and challenges; the identification of new problems and challenges; or the formulation of new theories, concepts, models or methods. We will then briefly revisit our case study results in order to illustrate what kinds of new knowledge we have been able to achieve, using SIM and CEM to investigate strategies used by designers when communicating the system's basic functions.

Regarding a new account of known problems, we can mention the set of results associated with the fact that users often miss the underlying design rationale and engage in sub-optimal (if not faulty) interaction. Even if help systems contain instructions and information leading to more productive interaction, users typically do not read it. The problem has been known for over two decades. Carroll and Rosson (1987) discussed it with respect to *the paradox of the active user*. The novelty brought about by semiotic engineering is to characterize this problem not as a matter of *providing agile access to help and instructions*, but of integrating the three classes of signs—metalinguistic, static and dynamic—into the interface layout, control and behavior, in such a way that messages conveyed by each constantly reinforce, evoke, complement and refer to each other. This can guide the users' *semiosis* towards the interpretations anticipated by designers, helping them to *infer* the design rationale from interaction itself. This also illustrates how new accounts of known problems naturally entail new kinds of solutions.

Regarding the identification of new solutions, partial or complete, generic or specific, to known problems and challenges, we can illustrate our contribution with the challenge of meeting the needs and expectations of a wide scope of intended users. This is a well-known problem in HCI, whose solution we proposed to advance using alternative metacommunication templates for different user profiles. This is not the same kind of solution found in customizable interfaces, whose options typically refer to variations in task models (i.e., on how users prefer to accomplish their tasks) and interface appearance (e.g., alternative "skins," hiding and showing interface elements, and the like). We proposed to design and implement different conversations about the same piece of software for different kinds of users.

The same example can be used to illustrate the contribution of SIM and CEM for the identification of a new challenge. The design and implementation of different conversations about the same piece of software for different kinds of users brings up new challenges. As is the case with transformative research strategies (Cresswell, 2009), this kind of solution points to the need for *cross articulation*—specifically that between semiotic engineering and AI. As discussed in Chapter 4, we should examine how AI techniques used in intelligent question-answering systems might be combined with interaction models like MoLIC (Barbosa & Paula, 2003; Paula, 2007), allowing us to operationalize alternate metacommunication templates depending on user preferences expressed in dialogs like the one proposed in Figure 4.16.

As an example of how SIM and CEM can contribute to the identification of new problems we can mention what we consider the most unpredictable and motivating knowledge obtained from our case study. It refers to what we have called *the recursive metacommunication structure*, a new research topic that aims at investigating how metacommunication in the lower level of software structure affects metacommunication at the top level one—the interface. As discussed, FOSS and also commercial development practices frequently explore the use of components (e.g., frameworks and programming languages). An unexpected consequence of these practices is that metacommunication elements embedded in a particular component's behavior may have undesirable effects on the designers' metacommunication message encoded in the system's interface elements. In our analysis of this situation, programmers were viewed as users and programming as interaction (with software components, integrated development environments, and the like). This new research problem leads us to seek articulation between semiotic engineering and programming or/and software engineering.

Finally, regarding the formulation of new theories, concepts, models and methods the very concept of *recursive metacommunication structure*, mentioned above, can be used to illustrate the reach of SIM and CEM in in-depth studies like the one we carried out with Audacity. Along with it came the realization that we need to develop new semiotic engineering methods and tools to support *programmers* and *developers*. To date, our focus had been placed on system's *designers*, but this research clearly shows that the very implementation of semiotic engineering solutions must be supported by epistemic tools that will raise the developers' awareness of how certain implementation alternatives might introduce inconsistencies in the global metacommunication message of the system.

Last, but not least, we should include in a *lessons learned* chapter such as this one that, although our primary goal with this book is to show how semiotic engineering methods can be used in scientific research contexts, the reaction of Audacity Team members to our research results revealed that the notion that designers communicate with users through systems' interfaces resonates very clearly with many aspects of what designers and developers view themselves as doing. This is particularly interesting for a theory of HCI where, as mentioned in Chapter 2, unlike others, designers play a *first-person role* in its proposed model of HCI. So, we have now considerable evidence that in technical contexts of professional practice semiotic engineering concepts and methods can also promote new insights and ideas. In the next and last chapter of this book, we will present our vision of where semiotic engineering is now heading.

* * * *

CHAPTER 6

The Near Future of Semiotic Engineering

We began this book by drawing attention to the recent calls for innovation coming from specialists in HCI. We stressed their opinions about the importance of using and proposing different scientific methods to stimulate and support not only the development of creative interfaces and interaction paths but also, and mainly, the generation of innovative HCI knowledge (Shneiderman, 2007; Greenberg & Buxton, 2008). We now end the book talking about the promises of innovation that we see in semiotic engineering.

In HCI, as is the case with many other fields of research, there is a strong tendency to associate scientific methodology with the idea of replicability. Many believe that the only acceptable methodological procedure in scientific research is to test hypotheses, following a rigorous and well-defined sequence of steps that yields results that other researchers can reproduce by following the same procedure in similar circumstances. The quantum of creativity and innovation, in this perspective, is introduced with the hypotheses formulated by researchers and subsequent discussions of results, but not with the results of methods themselves. These say no more than if the hypotheses passed the test or not. Therefore, in this perspective, creativity and innovation are mainly associated with inspiration, individual talent, insights promoted by accidental and unexpected outcomes, and other factors that are not determined by systematic methodological procedures. In fact, creativity and innovation are usually placed at the opposite end of systematic practices.

These perceptions usually prevent researchers from examining, experimenting, and possibly adopting a different perspective on scientific research. In it, methods, which by necessity are also defined in terms of systematic steps, may support and even stimulate innovation and feed inspiration. Such is the case of methods used in qualitative research (Denzin & Lincoln, 2000; Cresswell, 2009), in which the rigorous (re)construction of meanings impregnated in empirical data is the essence of research activity.

When reporting the results of a long-term empirical study with 12 innovative engineering companies, Petre (2004) underlined the importance of "systematic practices" in fostering innovation. In her words:

> Although the "Eureka!" effect has its place in radical innovation in engineering, and many design engineers experience moments of inspiration during design, innovation is more often incremental than radical. (. . .) in high-performing engineering teams and companies, innovation happens deliberately and moreover (. . .) such teams have developed a number of systematic practices that support innovation and feed inspiration (Petre, 2004, p. 447).

Notice the author's distinction between inspiration, a radical and occasional experience, and innovation, an incremental, deliberate, and systematic routine. In Petre's study, some systematic practices were identified. Most of them are related to broadening or changing an individual's view of the problem or of what might be a solution. Moreover, her study also showed that the professionals involved in these kinds of practices are "reflective practitioners" (Schön, 1983), engaged in constant analytical evaluations of their own work.

Quoting Schön's definition, Petre's work makes an important association between innovation and epistemic practices, an association that sheds light on important aspects of our own discussions in this book.

Methods that aim at replicating knowledge and producing generalizations rather than exploring open meaning possibilities, that is, predictive methods, are not likely to nurture innovation practices. In fact, the aim to produce "replicable ready-to-use solutions" frequently limits the interpretive and reflective processes of those who apply such methods, be they designers or researchers (see Greenberg & Buxton, 2008).

Nonpredictive methods typically used in qualitative and exploratory research, however, have different goals. They aim at producing precisely the kinds of practices and reflection identified in Petre's study with innovative companies. They aim to broaden or change the individual's view of problems (and/or solutions) pertaining to a certain object, phenomenon, or domain of activity. They actually support and stimulate semiotic processes discussed in this book, because they examine and explore the evolution of meanings associated to problems and solutions.

Although, just like predictive ones, nonpredictive methods are defined as a rigorous and well-defined sequence of steps, they are different because they lead researchers into unknown domains and problems, serving as reliable epistemic tools (Schön, 1983). They support and stimulate abductive reasoning. In this type of reasoning, as mentioned in previous chapters, plausible hypotheses generated with only partial evidence are taken as explanatory principles considered true until counterfactual evidence is found. In this case, the principle is revised, generating new hypotheses and explanatory principles, considered true until new counterfactual evidence is found, etc.

We should recall, at this point, that abductive reasoning is one of Peirce's (1992–1998) main contributions to logic and philosophy. It constitutes one of the pillars of a long and deeply rooted tradition in semiotics, whose followers argue convincingly that this kind of reasoning is the only one (compared to deduction and induction) that legitimately introduces new knowledge in the reasoner's knowledge base (Santaella, 2004).

It should thus be clear that semiotic engineering and its methods, defined as epistemic tools (de Souza, 2005), are well suited for innovation-oriented research. As the case study with Audacity has shown, the theory can be useful in scientific research as well as in professional design and development contexts. Our qualitative approach is a promising alternative to explore innovative framings and interpretations of HCI problems, adding new possibilities to knowledge generation backed up mainly by experimental methods and hypothesis testing.

However, the computer science community has been mainly interested in the predictive power of knowledge produced by scientific research. In this sense, semiotic engineering and its methods are not likely to be quickly assimilated by researchers that are trained and proficient in predictive science. It may take some time and critical mass for this community to begin to accept that, although designed to investigate singular concrete instances of objects and phenomena, qualitative methods can and should produce knowledge abstractions that are as valid as those produced by quantitative and experimental methods.

The critical difference is that such abstractions must not be used to predict other instances of objects and phenomena. They can be used to understand and explain them, to approach them and examine them, and even to influence them, bearing in mind that to influence is not to cause. They can be used as "systematic practices that support innovation and feed inspiration" (Petre, 2004, p. 447).

So in the near future of semiotic engineering, we see researchers beginning to explore innovative power of the theory. On the methodological front, they may try to expand and refine knowledge about other methods and theories that can be combined with ours. In particular, we think that they should try to see if and how quantitative methods can be used in triangulation with SIM and CEM. Another promising avenue is to explore how designers can use different types interaction diagrams to express their metacommunication intent and use these diagrams to analyze further aspects of the reception of the designers' message that we currently cannot examine. This would be an important methodological innovation for the theory itself and would reveal aspects of HCI that have not been empirically explored up to now. Finally, we believe that an inspection of the recursive metacommunication structure in software development is one of the most promising and revolutionary items in our own research agenda. Nevertheless, this kind of research is certainly a long-term project, whose results and impact are not likely to be well understood and fairly assessed in the near future.

The future calls for innovation. We hope this book will inspire innovative HCI researchers and practitioners to engage in the continuously unfolding semiosic process that once lured our imagination. Semiotic engineering has achieved important results, but much more is yet to be done. When we begin to view software as a collection of intentional signs weaving an intricate web of computer-mediated human communications, *everything changes*. It is like a twist in a kaleidoscope, the whole landscape is suddenly different, and in some ways, it will never be the same as before. A whole new story waits to be told.

* * * *

Acknowledgments

We are thankful to many people that helped us shape and improve this book in many different ways. We hope to have the chance to show them our gratitude in person. The names we mention here represent many others. We thank John Carroll for encouraging us to write the book in the first place. We owe many thanks to Robin Laffon, who introduced us to Audacity, and to Vaughan Johnson, who put us in contact with the Audacity Team and allowed us to carry out one of the richest research studies we have ever done. We are grateful to the members of the Audacity Team and the test participants, whose cooperation made this project possible. We also owe much to the 2008 class of our Semiotic Engineering Course at PUC-Rio—Andréia, Chantal, Laufer, Luciana, Márcio, Sílvia, Thaís, Tomás, and Ugo. Laufer and Ugo, in particular, kindly allowed us to mention parts of their projects in this book. We thank Daniel Schwabe, Simone Barbosa, Bruno Santana, and Maíra de Paula for their comments on the manuscript. The final text also incorporates the suggestions of editorial reviewers, for which we are especially thankful. Last, but not least, we would like to thank Diane Cerra for helping us through the whole publishing process.

The Department of Informatics at PUC-Rio provided us with excellent infrastructure and intellectual stimulus while we were preparing the manuscript. Clarisse de Souza had additional support from CNPq, the Brazilian Council for Scientific and Technological Development.

Writing a book is a highly absorbing activity, and our families and friends often missed, quietly, some of our love and attention. Their unconditional love and support, cheering us all the way to this happy end, infused us with the essential ingredients of fulfilling achievements. We would never be able to thank them enough for that.

Glossary

Abduction A method of logical reasoning also known as abductive reasoning or hypothetical reasoning. In it, plausible hypotheses built from partial evidence provided by readily observable phenomena are "precipitously" generalized in the form of a principle or rule that *explains* a certain number of facts that call for meaning or explanation. They are held as *true* until counterfactual evidence is found, in which case they are revised or replaced by more plausible ones (till new counterfactual evidence is found). In semiotic theories of Peircean extraction, this recursive self-correcting form of reasoning describes the continuous process of interpretation by which meaning is assigned to potentially significant observations. pp. 17, 76, 106–107.

Communicability A system's (i.e., the designer's deputy's) capacity to achieve full metacommunication through its interface. During interaction, it must communicate to users, in an organized and resourceful way, the underlying design rationale and interactive principles that constitute the designer's vision of his design product. pp. 24–25.

Communication The process through which the possibilities of various signification systems are explored in order to generate expressions that are meant to achieve a very wide range of goals and effects. In an *ideal* communication, a sender produces signs to express his communicative intent to a receiver, who by capturing the sender's message behaves in such way that the sender's intent is achieved. pp. 14, 16–17.

Designers' deputy The communicating agent that tells the designer's message to users at interaction time, namely, the computer system itself, through its interface. The system is the designer's proxy. It is capable to express *all* and *only* the computationally encoded message contents elaborated by the designer at design time. pp. 17, 24–25.

Discourse (interactive discourse) The computational version of the conversations that designers would have with users in order to achieve the ultimate communicative intent in design. Because designers are not present at interaction time (and are represented by the system itself), all the interactive discourse must have been planned for at design time and implemented in the form of a computer program in subsequent development stages. pp. 10, 17, 86–87.

Dynamic signs Interface signs that emerge with interaction and must be interpreted with reference to it. Dynamic signs are bound to temporal and causal aspects of the interface and communicate the processing that leads to transitions between system states. p. 19.

Epistemic tools Tools that are not used to yield directly the answer to the problem, but to increase the problem-solver's understanding of the problem itself and the implications it brings about. pp. 15, 23, 106–107.

Epistemology The study of the nature, origin, and scope of knowledge and of what there is to be known and how such knowledge originates, expands, and collapses. pp. 20, 90.

Interlocutor Each one of the speakers and listeners involved in a communication process. pp. 10, 20, 76.

Metacommunication Communication about (aspects of) communication itself. In semiotic engineering, the concept of metacommunication refers to the idea that, through interface elements and interaction patterns, designers communicate their design vision to users, telling them how, why, when, and what for to interact with the very system they are using. p. 13.

Metalinguistic signs Signs that explicitly inform, illustrate, or explain the meaning of static and dynamic signs. They come in the form of help or error messages, warnings, clarification dialogs, tips, and the like. With metalinguistic signs, designers explicitly communicate to users the meanings encoded in the system and how they can be used. pp. 19–20.

Metaphor A figure of speech in which a sign that ordinarily designates one thing is used to designate another (similar) one. An analogy is made between two seemingly unrelated *objects* because they have some important quality in common (e.g., "good research is a jewel"). pp. 5, 28.

Metonymy A figure of speech in which a sign designates another due its semantic contiguity (or "relatedness"). Some popular forms of metonymy designate objects by their brand (e.g., "drink Budweiser"), institutions by their location (e.g., "Rome condemned the archbishop"), people by their appearance (e.g., "the blue eyes are back"), etc. p. 24.

Semiosis The process of sign interpretation that leads to the continuous production of meaning. Semiosis is the theoretical concept corresponding to *sense making*, whose logical counterpart is abductive reasoning (see above). pp. 17–18, 20, 76, 86, 97.

Semiotics A heterogeneous theoretical field that investigates signs and signification processes. Semiotics also involves the study of how signs and signification take part in communication, as well as how signs relate to culture. pp. 13–14, 17, 20, 107.

Sign In Peircean semiotics, a sign is anything that somebody takes to stand for something else in some respect or capacity. Nothing is a sign unless (or until) it is interpreted by somebody. The same signs may have very different and equally legitimate meanings for different people, in different contexts. Likewise, that which is a sign for one individual is not necessarily a sign for another. pp. 7, 14, 17.

Signification The process through which certain systems of signs are established by virtue of social and cultural conventions adopted by the interpreters and producers of such signs. The product of this process is a systematic and culturally motivated set of associations between contents and expressions. pp. 2, 10, 14, 20–21.

Static signs Interface signs whose meaning is interpreted independently of temporal and causal relations (e.g. press buttons, pull down menus, and the like). Static signs stimulate the user to engage in interaction with the artifact and to anticipate what the interaction would be like and what consequences it should bring about. p. 19.

Triangulation The last methodological procedure in qualitative research required to ensure the scientific validity of achieved results. Triangulation may be done by comparing data achieved by different methods, by involving different researchers in applying the same method, or by using multiple theoretical perspectives to interpret the data. The purpose of triangulation is not to replicate results, but to verify that all results are consistent with each other. pp. 27, 33, 48, 101.

References

Andersen, P. B. (1997). *A theory of computer semiotics: Semiotic approaches to construction and assessment of computer systems* (2nd ed.). Cambridge: Cambridge University Press.

Andersen, P. B., Holmqvist, B., & Jensen, J. F. (Eds.). (1993). *The computer as medium.* Cambridge: Cambridge University Press.

Baranauskas, C. M., Salles, J. P., & Liu, K. (2003). Analysing communication in the context of a software production organisation. In M. Piattini, J. Filipe, & J. Braz (Eds.), *Enterprise information systems IV* (pp. 202–209). Hingham, MA: Kluwer Academic Publishers.

Barbosa, C. M. A. (2006). *Manas: uma ferramenta epistêmica. De apoio ao projeto da comunicação em sistemas colaborativos.* (Ph.D. Thesis applied to Computer Science Department, PUC-Rio, Brazil). Retrieved from http://www2.dbd.puc-rio.br/pergamum/biblioteca/php/mostrateses.php?open=1&arqtese=0210647_06_Indice.html. Last visited in October 2008.

Barbosa, C. M. A., Leitão, C. F., & de Souza, C. S. (2005). Why understanding culture is important for developing effective online health support: The Brazilian context. Paper presented at *HCI International 2005*, Las Vegas, USA.

Barbosa, S. D. J., & de Souza, C. S. (2001). Extending software through metaphors and metonymies. *Knowledge Based Systems*, 14(1–2), pp. 15–27.

Barbosa, S. D. J., & Paula, M. G. (2003). Designing and evaluating interaction as conversation: A modeling language based on semiotic engineering. In J. Jorge, N. J. Nunes, & J. Falcão e Cunha (Eds.), *Interactive systems design, specification, and verification—10th International Workshop, DSV-IS 2003* (pp. 16–33). Madeira Island, Portugal: Lecture Notes in Computer Science.

Barcellini, F., Détienne, F., & Burkhardt, J. M. (2007). Cross-participants: fostering design-use mediation in an open source software community. In *Proceedings of the 14th European Conference on Cognitive Ergonomics: Invent! Explore!* (pp. 57–64). London, United Kingdom: ACM.

Bentkowksa-Kafel, A., Cashen, T., & Gardiner, H. (Eds.). (2008). *Futures past: Thirty years of arts computing.* Oxford, UK: Intellect.

Beyer, H., & Holtzblatt, K. (1998). *Contextual design: Defining customer-centered systems.* San Francisco: Morgan Kaufmann.

Carroll, J. M. (2000). *Making use: Scenario-based design of human–computer interactions*. Cambridge, MA: MIT Press.

Carroll, J. M. (2003). *HCI models, theories, and frameworks: Toward a multidisciplinary science*. Menlo Park: Morgan Kaufmann.

Carroll, J. M., & Rosson, M. B. (1987). The paradox of the active user. In J. M. Carroll (Ed.), *Interfacing thought: Cognitive aspects of human–computer interaction* (pp. 26–28). Cambridge, MA: MIT Press.

Cooper, G., Hine, C., Rachel, J., & Woolgar, S. (1995). Ethnography and human–computer interaction. In P. J. Thomas (Ed.), *The social and interactional dimensions of human–computer interfaces* (pp. 11–36). New York, NY: Cambridge Series On Human–Computer Interaction, Cambridge University Press.

Cresswell, J. W. (2009). *Research design: Qualitative, quantitative, and mixed methods approaches* (2nd ed.). Thousand Oaks: Sage Publications.

Cunha, C. K. V. (2001). *Um Modelo Semiótico dos Processos de Comunicação Relacionados à Atividade de Extensão à Aplicação por Usuários Finais*. (Ph.D. Thesis applied to Computer Science Department, PUC-Rio, Brazil). Retrieved from http://www.maxwell.lambda.ele.puc-rio.br/cgi-bin/db2www/PRG_0651.D2W/SHOW?Mat=&Sys=&Nr=&Fun=&CdLinPrg=pt&Cont=6520:pt. Last visited in October 2008.

da Silva, S. R. P. (2001). *Um Modelo Semiótico para Programação por Usuários Finais*. (Ph.D. Thesis applied to Computer Science Department, PUC-Rio, Brazil).

de Souza, C. S. (1993). The semiotic engineering of user interface languages. *International Journal of Man–Machine Studies*, 39(4), pp. 753–773.

de Souza, C. S. (2005). *The semiotic engineering of human–computer interaction*. Cambridge, MA: MIT Press.

de Souza, C. S., & Cypher, A. (2008). Semiotic engineering in practice: redesigning the CoScripter interface. In: *Proceedings of the Working Conference on Advanced Visual interfaces* (pp. 165–172). New York, NY: ACM.

de Souza, C. S., & Preece, J. (2004). A framework for analyzing and understanding online communities. *Interacting with Computers*, 16(3), pp. 579–610.

de Souza, C. S., & Sedig, K. (2001). Semiotic considerations on direct concept manipulation as a distinct interface style for learnware. In *Proceedings of IV Brazilian Workshop on Human Factors in Computing Systems* (pp. 229–241). Porto Alegre, Brasil: SBC.

de Souza, C. S., Laffon, R. F., & Leitão, C. F. (2008). Communicability in multicultural contexts: A study with the International Children's Digital Library. In *Proceedings of Human–Computer Interaction Symposium—HCIS 2008* (pp. 129–142). Boston: IFIP-Springer.

de Souza, C. S., Leitão, C. F., Prates, R. O., Bim, S. A., & da Silva, E. J. (to appear). *Using the semiotic inspection method in scientific research contexts.*

de Souza, C. S., Leitão, C. F., Prates, R. O., & da Silva, E. J. (2006). The semiotic inspection method. In *Proceedings of the 7th Brazilian Symposium of Human Factors on Computer Systems* (pp. 148–157). Porto Alegre, Brasil: SBC.

de Souza, C. S., Nicolaci-da-Costa, A. M., da Silva, E. J., & Prates, R. O. (2004). Compulsory institutionalization: Investigating the paradox of computer-supported informal social processes. *Interacting With Computers*, 6(4), pp. 635–656.

de Souza, C. S., Prates, R. O., & Carey, T. (2000). Missing and declining affordances: Are these appropriate concepts? *Journal of the Brazilian Computer Society*, 1(7), pp. 26–34.

Denzin, N. K., & Lincoln, Y. S. (Eds.). (2000). *Handbook of qualitative research* (2nd ed.). Thousand Oaks: Sage Publications.

Dourish, P. (2001). *Where the action is.* Cambridge, MA: The MIT Press.

Eco, U. (1976). *A theory of semiotics.* Bloomington, IN: Indiana University Press.

Erickson, T. (1995). Notes on design practice: Stories and prototypes as catalysts for communication. In J. M. Carroll (Ed.), *Scenario-based design: Envisioning work and technology in system development* (pp. 37–58). New York, NY: John Wiley & Sons.

Fogg, B. J. (2003). *Persuasive technology.* Menlo Park, CA: Morgan Kaufmann.

Gelernter, J., & Jagganathan, S. (1990). *Programming linguistics.* Cambridge, MA: The MIT Press.

Greenberg, S., & Buxton, B. (2008). Usability evaluation considered harmful (some of the time). In *Proceedings of ACM CHI 2008 Conference on Human Factors in Computing Systems* (pp. 111–120). Florence, Italy.

Jorna, R. J. (1990). Knowledge representation and symbols in the mind: An analysis of the notion of representation and symbol in cognitive psychology. Tübingen: Stauffenburg-Verlag.

Jorna, R. J. (1994). *Semiotic engineering: modeling information systems by signs.* Unpublished manuscript presented at the Fifth Congress of the International Association for Semiotic Studies, Berkeley, USA, June, 12–18, 1994.

Jorna, R. J., & van Wezel, W. (1995). Worldmaking with objects: A case in semiotic engineering. *SOM Research Report 95A12.* Groningen: University of Groningen. Retrieved from http://irs.ub.rug.nl/ppn/143360949. Last visited in October 2008.

Kammersgaard, J. (1988). Four different perspectives on human–computer interaction. *International Journal of Man–Machine Studies*, 28(4), pp. 343–362.

Kobsa, A., & Wahlster, W. (Eds.). (1990). *User models in dialog systems.* New York: Springer-Verlag New York, Inc.

Leech, G. (1983). *The principles of pragmatics.* London: Longman.

Leitão, C. F., de Souza, C. S., & Barbosa, C. M. A. (2007). Face-to-face sociability signs made explicit in CMC. In *Proceedings of Interact 2007, 11th IFIP TC 13 International Conference* (pp. 5–18). Berlin, Heidelberg.

Leite, J. C. (1998). *Modelos e formalismos para a engenharia semiótica de interfaces de usuário* (Ph.D. Thesis applied to Computer Science Department, PUC-Rio, Brazil). Retrieved from http://www.maxwell.lambda.ele.puc-rio.br/cgi-bin/db2www/PRG_0651.D2W/SHOW?Mat=&Sys=&Nr=&Fun=&CdLinPrg=pt&Cont=1809:pt. Last visited in October 2008.

Lieberman, H., Paternò, F., & Wulf, V. (Eds.). (2006). *End-user development*. Dordrecht: Springer.

Maybury, M. T. (Ed.). (1993). *Intelligent multimedia interfaces*. Cambridge, MA: The MIT Press.

Maybury, M. T. (Ed.). (2004). *New directions in question answering*. Cambridge, MA: The MIT Press.

Moore, J. (1995). *Participating in explanatory dialogues*. Cambridge, MA: MIT Press.

Muller, M. J., & Kuhn, S. (1993). Participatory design. *Commun. ACM*, 36(6), pp. 24–28.

Nadin, M. (1988). Interface design and evaluation. In R. Hartson & D. Hix (Eds.), *Advances in human–computer interaction* (Vol. 2). Norwood, NJ: Ablex Publishing Corp.

Nardi, B. A. (1996). Context and consciousness: Activity theory and human–computer interaction. Cambridge, MA: MIT Press.

Nichols, D. M., & Twidale, M. B. (2003). The usability of open source software. *First Monday*, 8(1). Retrieved from http://firstmonday.org/issues/issue8_1/nichols/. Last visited in October 2008.

Nielsen, J. (1994). Heuristic evaluation. In J. Nielsen & R. L. Mack (Eds.), *Usability inspection methods* (pp. 25–62). New York, NY: John Wiley.

Norman, D. A. (n.d.). Design as communication. Retrieved from: http://jnd.org/dn.mss/design_as_communication.html. Last visited in October 2008.

Norman, D. A. (1986). Cognitive engineering. In D. A. Norman & S. W. Draper. *User-centered system design* (pp. 31–62). Hillsdale, NJ: Laurence Erlbaum.

Norman, D. A. (2007). *The design of future things*. New York: Basic Books.

Patton, M. Q. (2001). *Qualitative evaluation and research methods* (2nd ed.). Thousand Oaks: Sage Publications.

Paula, M. G. (2007). *ComunIHC-ES: ferramenta de apoio à comunicação entre profissionais de IHC e engenheiros de software*. (Ph.D. Thesis applied to Computer Science Department, PUC-Rio, Brazil). Retrieved from http://www2.dbd.puc-rio.br/pergamum/biblioteca/php/mostrateses.php?open=1&arqtese=0310850_07_Indice.html. Last visited in October 2008.

Peirce, C. S. (1992–1998). In N. Houser & C. Kloesel (Eds.), *The essential Peirce* (Vols. I and II). Bloomington, IN: Indiana University Press.

Petre, M. (2004). How expert engineering teams use disciplines of innovation. *Design Studies*, 25(5), pp. 477–493.

Prates, R. O. (1998). *A engenharia semiótica de linguagens de interfaces multi-usuário*. (Ph.D. Thesis applied to Computer Science Department, PUC-Rio, Brazil). Retrieved from http://www.maxwell.lambda.ele.puc-rio.br/cgi-bin/db2www/PRG_0651.D2W/SHOW?Mat=&Sys=&Nr=&Fun=&CdLinPrg=pt&Cont=1813:pt. Last visited in October 2008.

Prates, R. O., Barbosa, S. D. J., & de Souza, C. S. (2000) A case study for evaluating interface design through communicability. *International Conference on Designing Interactive Systems—DIS2000* (pp. 308–317). New York: ACM Press.

Prates, R. O., de Souza, C. S., & Barbosa, S. D. J. (2000). A method for evaluating the communicability of user interfaces. *ACM Interactions*, 7(1), 31–38.

Reeves, B., & Nass, C. (1996). The media equation: How people treat computers, television and new media like real people and places. Cambridge: Cambridge University Press.

Rheinfrank, J., & Evenson, S. (1996). Design languages. In T. Winograd (Ed.), *Bringing design to software* (pp. 63–85). Reading, MA: Addison-Wesley.

Santaella, L. B. (2004). *O método anti-cartesiano de C. S. Peirce*. São Paulo, Brasil: UNESP/FAPESP.

Schön, D. A. (1983). *The reflective practitioner: How professionals think in action*. New York: Basic Books.

Schön, D., & Bennett, J. (1996). Reflective conversation with materials. In T. Winograd (Ed.), *Bringing design to software* (pp. 171–184). Reading, MA: Addison-Wesley.

Searle, J. R. (1979). *Expression and meaning*. Cambridge: Cambridge University Press.

Seidman, I. (1998). Interviewing as qualitative research: A guide for researchers in education and the social sciences. New York: Teachers College Press.

Sharp, H., Rogers, Y., & Preece, J. (2007). *Interaction design: Beyond human computer interaction*. Hoboken, NJ: Wiley. Online at http://www.id-book.com.

Shneiderman, B. (2007). Creativity support tools: Accelerating discovery and innovation. *Commun. ACM*, 50(12), 20–32.

Shneiderman, B., & Plaisant, C. (2004). Designing the user interface: Strategies for effective human–computer interaction. Reading, MA: Addison-Wesley.

Silva, B. S., & Barbosa, S. D. J. (2007). Designing human–computer interaction with MoLIC diagrams—A practical guide. In C. J. P. de Lucena (Ed.), *Monografias em Ciência da Computação*. (*PUC-Rio Inf MCC12/07*. Computer Science Department, PUC-Rio, Brazil). Retrieved from: http://www-di.inf.puc-rio.br/~simone/publications/07_12_silva.pdf. Last visited in October 2008.

Silveira, M. S. (2002). *Metacomunicação designer-usuário na interação humano-computador*. (Ph.D. Thesis applied to Computer Science Department, PUC-Rio, Brazil). Retrieved from http://www.maxwell.lambda.ele.puc-rio.br/cgi-bin/db2www/PRG_0651.D2W/SHOW?Mat=&Sys=&Nr=&Fun=&CdLinPrg=pt&Cont=3920:pt. Last visited in October 2008.

Suchman, L. (1987). Plans and situated actions: The problem of human–machine communication. Cambridge: Cambridge University Press.

Sundar, S. S., & Nass, C. (2000). Source orientation in human–computer interaction. *Communication Research*, 27(6), pp. 683–703.

Twidale, M. B., & Nichols, D. M. (2005). Exploring usability discussions in open source development. In *Proceedings of the Proceedings of the 38th Annual Hawaii International Conference on System Sciences (HICSS'05)—Track 7*. Washington, DC: IEEE Computer Society. Retrieved from http://dx.doi.org/10.1109/HICSS.2005.266. Last visited in October 2008.

van Heusden, B., & Jorna, R. J.(2001). Toward a semiotic theory of cognitive dynamics in organizations. In K. Liu, R. J. Clarke, P. B. Andersen, & R. K. Stamper (Eds.), *Information, organisation and technology: Studies in organisational semiotics* (pp. 83–113). Boston, MA: Kluwer.

Wharton, C., Rieman, J., Lewis, C., & Polson, P. (1994). The cognitive walkthrough method: A practitioner's guide. In: J. Nielsen & R. Mack (Eds.), *Usability inspection methods* (pp. 105–140). NY: Wiley.

Winograd, T., & Flores, F. (1986). *Understanding computers and cognition.* New York: Addison-Wesley.

Author Biographies

Clarisse Sieckenius de Souza is a professor at the Department of Informatics of Pontifical Catholic University of Rio de Janeiro (PUC-Rio). With a Ph.D. in applied linguistics, she started in computer science doing research in natural language processing and text generation. She gradually moved from artificial intelligence into HCI, switching her interdisciplinary base from linguistics to semiotics. She is the founder and principle investigator of the Semiotic Engineering Research Group (SERG) at PUC-Rio, which produced the most comprehensive semiotic account of HCI in Informatics to date.

Carla Faria Leitão is a senior researcher of SERG at PUC-Rio. She has a Ph.D. in clinical psychology and has been drawn to informatics and HCI by the subjective aspects of computer artifacts exposed and explored by semiotic engineering. With a special interest in how qualitative methods can be used outside the social and human sciences, she has been leading methodological research in SERG since 2002.